Logistics Engineering

Edited by Samson Jerold Samuel Chelladurai, Suresh Mayilswamy, S. Gnanasekaran and Ramakrishnan Thirumalaisamy

Published in London, United Kingdom

Logistics Engineering
http://dx.doi.org/10.5772/intechopen.97960
Edited by Samson Jerold Samuel Chelladurai, Suresh Mayilswamy, S. Gnanasekaran and Ramakrishnan Thirumalaisamy

Contributors
Thomas E. Fernandez, M. Rameswari, Theophilus Chinonyerem Nwokedi, Baha M. Mohsen, Yonit Barron, Herbert Ruile, Lukas Lichtsteiner, Fraidoon Karimi, Sanja Bauk

First published in London, United Kingdom, 2022 by IntechOpen
IntechOpen is the global imprint of INTECHOPEN LIMITED, registered in England and Wales, registration number: 11086078, 5 Princes Gate Court, London, SW7 2QJ, United Kingdom

British Library Cataloguing-in-Publication Data
A catalogue record for this book is available from the British Library

Additional hard and PDF copies can be obtained from orders@intechopen.com

Logistics Engineering
Edited by Samson Jerold Samuel Chelladurai, Suresh Mayilswamy, S. Gnanasekaran and Ramakrishnan Thirumalaisamy
p. cm.
Print ISBN 978-1-80355-750-2
Online ISBN 978-1-80355-751-9
eBook (PDF) ISBN 978-1-80355-752-6

Meet the editors

Dr. Samson Jerold Samuel Chelladurai, Ph.D., is an associate professor in the Department of Mechanical Engineering, Sri Krishna College of Engineering and Technology (SKCET), India. He obtained a Ph.D. in Composite Materials from Anna University, India in 2018. He is specialized in the manufacturing of metal matrix composites, characterization of composites, tribology, and optimization techniques. He has published many journal articles and five books. He has published five patents and received grants worth $89,000 USD. He has received several awards from SKCETand other organizations.

Dr. Suresh Mayilswamy, Ph.D., is an associate professor in the Department of Robotics and Automation Engineering, PSG College of Technology, India. His Ph.D. research focused on theoretical and experimental investigations of orientations of brake pads in vibratory part feeders. His current research focuses on advanced manufacturing technology, process automation, and industrial robotics. He has published seventy-three technical papers in refereed journals and three books on industrial automation and assembly automation. He received the 2013 outstanding faculty award from Sri Krishna College of Engineering & Technology (SKCET), India, and the 2018 best faculty of the year award from PSG Institute of Advanced Studies, India.

Dr. S.Gnanasekaran, Ph.D., is an assistant professor in the Department of Mechanical Engineering, Sri Sakthi Institute of Engineering and Technology, India. He obtained a Ph.D. in Welding and Surface Engineering with research funded by the University Grants Commission - Department of Atomic Energy (UGC-DAE). He has published many journal articles and five patents. He has also published five book chapters. Dr. Gnanasekaran received several awards from international conferences for his technical work. He also received the best faculty award for his immeasurable contribution to students' welfare. He has been an editorial and advisory board member as well as a reviewer for several journals.

Dr. Ramakrishnan Thirumalaisamy is an associate professor in the Department of Mechanical Engineering, Sri Eshwar College of Engineering, India. He obtained a Ph.D. in Polymer Composites from Anna University, India in 2018. His areas of interest include biomaterials, materials design, tribology, and composite materials. He has published many journal articles and several patents. He has been an editorial and advisory board member as well as a reviewer for several journals. He is also a lifetime member of various scientific and professional bodies. Dr. Thirumalaisamy received a best researcher award and recognition for best paper from his institution, Sri Eshwar College of Engineering.

Contents

Preface

The book presents the key themes of logistics engineering, including supply chain management, logistics, and inventory control. It presents recent advances and developments in the field and discusses emerging research topics on blockchain technology in supply chain management, the safety of food in the supply chain, the environment of maritime transport, sustainable logistics, the autonomous warehouse, and fluid inventory control.

Samson Jerold Samuel Chelladurai
Associate Professor,
Mechanical Engineering,
Sri Krishna College of Engineering and Technology,
Coimbatore, Tamil Nadu, India

Suresh Mayilswamy
PSG College of Technology,
Coimbatore, India

S. Gnanasekaran
Sri Shakthi Institute of Engineering and Technology,
Tamil Nadu, India

Ramakrishnan Thirumalaisamy
Sri Eshwar College of Engineering,
Tamil Nadu, India

Section 1

Supply Chain

Chapter 1

Blockchain Technology in Supply Chain Management

Thomas E. Fernandez

Abstract

Supply chain management has existed for thousands of years. Technology started evolving with the industrial revolution and yet, paper-based bills of lading are still widely used in the international trade today. This chapter explores what problems exist in today's supply chains, and whether blockchain technology can help solve them. The problems are categorized into three categories: The origin of the raw material or the product is not obvious; trust issues can exist between seller and buyer; and the supply chain execution is inefficient. A brief explanation of blockchain technology follows as the second part, with a focus on the technology that is useful for this purpose of this article. The two parts come together in the third part, showing how blockchain technology can solve the problems. Finally, a critical view is taken to show the limitations of blockchain technology as a problem solver.

Keywords: blockchain technology, supply chains, supply chain management, use cases

1. Introduction

Supply chain management manages the flow of material, money, and information from the origin of the raw materials to the final consumer. Supply chains have existed for thousands of years, which includes procurement, transportation, inventory and warehouse management, customs clearance, and distribution. A supply chain involves many entities, including companies and governments, and several departments in each of these entities. However, 62% of the surveyed companies responded that they have only limited visibility of their supply chains [1].

In the latter part of the twentieth century, companies and governments started to digitize the management of the supply chain (also see [2]), first in technological silos in which each department had their own hardware and software, such as an inventory management system (IMS) or an accounting system, for example. With increasing acceptance of computers in businesses, an increasing amount of software entered the market. Examples are distribution and fleet management systems, automatic ordering systems, automatic storage and retrieval systems (AS/ARs) and transport management system (TMS) to name but a few. All these systems were finally integrated into an enterprise resource planning (ERP) system. However, not all ERP systems could integrate all the different software systems, and—most importantly—often not across companies within the same supply chain.

Vertical and horizontal integration requires the exchange of fast and accurate information between the various members of the supply chain. These members include first-tier and second-tier suppliers, wholesalers and retailers, as well as service providers [3]. Even without fully integrating the supply chains, information exchange is crucial to an efficient operation. The traditional method for an exporter was to type their invoice and packing list, send it to their forwarder, and have the forwarder retype it as a shipping instruction to the shipping line who in turn retyped the data into the bill of lading (B/L); and at destination, a customs broker retyped all the data into the customs clearance system. This is not only inefficient but can cause problems, starting from typing mistakes to the opportunity for fraud.

Blockchain technology is a technology that has many use cases. It became known for cryptocurrency, as the first cryptocurrency, Bitcoin, was also the first use case for blockchain technology [4]. This chapter will show how blockchain technology is much more versatile and can be used in supply chains in order to manage complex supply chains efficiently and accurately.

This chapter is structured as follows:

Section 1: Introduction.

Section 2: Problems intrinsic to traditional supply chains.

Section 3: Brief introduction to blockchain technology.

Section 4: How blockchain technology can help solve the problems.

Section 5: Possible problems with blockchains.

Section 6: Conclusion.

2. Problems intrinsic to traditional supply chains

Supply chain management manages the flow of material, money, and information. In this chapter, we will concentrate only on the material and the information flows. Material usually flows from the origin of the raw material to the consumer (downstream) and information flows both upstream and downstream in the supply chain. A simple supply chain showing the flow of material is shown in **Figure 1**.

With the advent of computers, supply chain visibility [5] was introduced. Supply chain visibility means that the supply chain manager can see—on his or her computer screen—where the inbound or outbound material is located at the very moment, whether stocks are getting low and need to be reordered, and where any problems exist. Airfreight shipments or seafreight containers can be tracked on the Internet, and warehouse and inventory management systems can be accessed remotely.

The following problems can be identified in current supply chains:

- The origin of the raw material or the product is not obvious.
- Trust issues can exist between seller and buyer.
- The supply chain execution is inefficient.

2.1 Origin of the raw material or the product

While the consumers in the past century often chose a product based on the price and the perceived quality—using the brand name and the retailer's brand image as references—today the consumer wants to know the origin of the goods. For example,

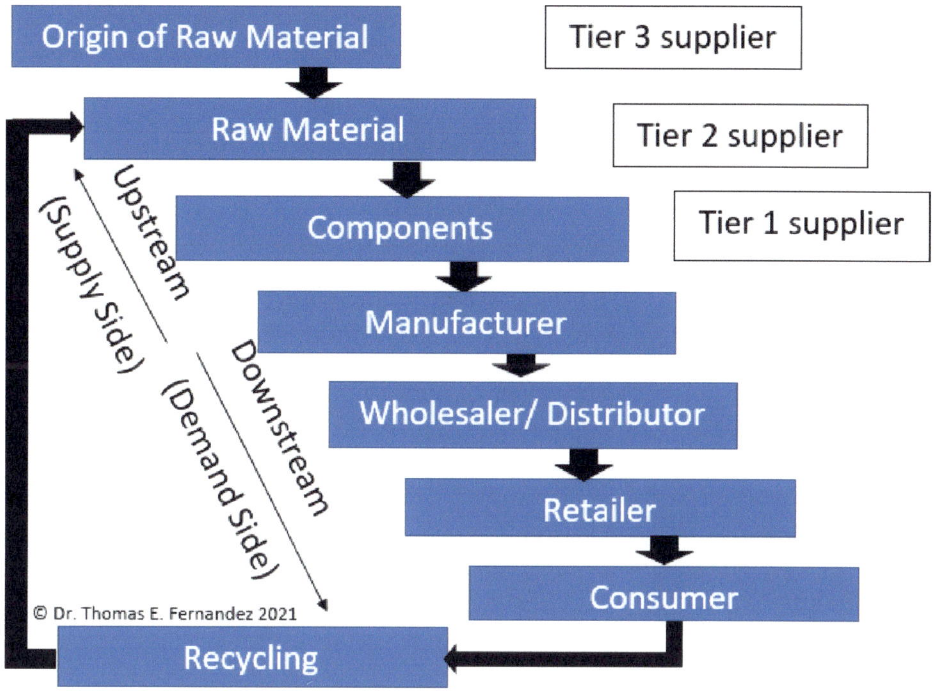

Figure 1.
A simple supply chain. Source: The author.

organic food is becoming more and more popular and so is the desire of the consumers to be able to check that the goods are really organic [6, 7]. Foodstuff has been falsely declared as organic in order to achieve higher prices. Another example is that dolphins are believed to be accidentally killed when fishing for tuna (see e.g., [8]), clothes may have been produced in sweat shops or using child labor [9] or that a wedding ring could contain a "conflict diamond" or "blood diamond" [10]. A scandal in Europe in 2013 revealed that lasagna contained horse meat, which was unexpected and undesired by the consumers. Lack of traceability was blamed [11].

The current tracking systems cannot provide the visibility needed to ensure that the goods come from an approved source. While seafreight containers can be traced from the port of loading or an airfreight shipment from the airport of departure, it is unclear where the goods actually originated from. Trucks, for example within Europe, can be tracked with GPS systems, but the shipments themselves cannot.

Suppliers may even intentionally falsify the origin of goods. This happens when the origin is illicit (e.g., conflict diamonds), the manufacturer handled unethically and should not be revealed (e.g., using child labor and forced labor), or an importer wants to take advantage of reduced import duties under a Free Trade Agreement (FTA) even though the goods do not satisfy the rules of origin specified in the FTA.

2.2 Trust issues between seller and buyer

Unless trust has already been established between seller and buyer, the seller needs to ensure that they will receive payment when they deliver the goods, while the buyer needs to ensure that they will receive the goods when they arrange for payment.

The solution has been for centuries that the buyer opens am Irrevocable Documentary Letter of Credit with their bank, which the seller accepts. This method requires a number of documents and is rather complex and inefficient, as well as costly. In addition, it bears some risks [12].

If the buyer requires goods or components made in a certain country as they believe the quality is better, they will just have to believe that the seller does not source these items somewhere else for a lower price.

2.3 Inefficient supply chain execution

Contracts may stipulate that certain actions will take place when certain conditions are met. For example, payment shall be affected when goods have been shipped, or when goods have been delivered. The problem here is that it requires human action, and if the conditions have been met and the human action has not been taken, there is little the beneficiary of the action (e.g., the recipient of the payment) can do. For large financial amounts, international legal action can be taken, but they are very costly and can take years.

If conditions trigger an automated action by a computer system, a risk of hacking exists. It can be forged that the conditions have been met so that action is taken.

3. Blockchain technology

This chapter provides a brief explanation of blockchain technology. Only aspects relevant to this article are described, and they are described with relevance to supply chains in mind.

Blockchain technology is a layer on the internet. It consists of "blocks" that are chained together. A block contains sets of data, similar to records in a database. While a block contains several sets of data, this chapter simplifies the technology into assuming each block contains one set of data in order to explain the principle.

A set of data contains

- Data (the content);
- A timestamp showing when it was created;
- The "hash" of the previous block;
- Its own "hash";
- Some other technical data.

A new database record will be created each time a transaction happens. In supply chains, an example of a transaction is loading of the container onto a ship.

3.1 What is a hash?

A hash is the output of an algorithm. Simply put, an algorithm is a mathematical formula consisting of more than one step. It calculates on output from one or more inputs. The output of the hash algorithm is a code, an encryption of the input. If we

INPUT	HASH
This is a test	C7BE1ED902FB8DD4D48997C6452F5D7E509FBCDBE2808B16BCF4EDCE4C07D14E
This is a test	2E99758548972A8E8822AD47FA1017FF72F06F3FF6A016851F45C398732BC50C

Source: Rosic (2020).

Table 1.
Changing "T" at the beginning to "t" [13].

INPUT	HASH
Hi	639EFCD08ABB273B1619E82E78C29A7DF02C1051B1820E99FC395DCAA3326B8
Welcome	53A53FC9E2A03F9B6E66D84BA701574CD9CF5F01FB498C41731881BCDC68A7C8

Source: Rosic (2020).

Table 2.
Length of input [13].

run the same input through the same algorithm several times, the output (hash) will always be the same. That is why it is sometimes called the fingerprint of the data. If only one letter is changed, the hash completely changes (**Table 1**).

On the other hand, it is practically impossible to re-engineer a hash and get the original input. If only the hash is known, the data that were supplied to the algorithm for calculation of this hash cannot be found out.

Furthermore, a hash has a fixed length. If the SHA256 algorithm is used, the length is 32 bytes citation regardless of how long the input was (**Table 2**):

3.2 Creating a chain

It was mentioned earlier that a data record in the blockchain contains the hash of the previous block. It is also known that the hash changes when any of the inputs change. It follows that the hash of the previous block cannot be changed without changing the hash of the current block; the current block is hence linked to the previous block. So, we have a chain of blocks, or better: a blockchain.

- Block1 → Block2 → Block3 → Block4

If each new block contains the hash of the previous block, this means that if a hacker wants to change any data in any previous block of the blockchain, this hacker will have to change the previous hashes in all previous blocks. If they want to change data in Block2, this changes the hash of Block2 and therefore the hash of the previous block for Block3. Hence, the hash of Block3 also changes, and the hacker needs to regenerate that has, and this continues along the whole blockchain to the end.

3.3 Distributed ledger technology

Once a hacker in the earlier-mentioned scenario has access to the database, it would not be a problem to change all the hashes. This is where distributed ledger technology (DLT) comes in. A ledger is a type of database. If DLT technology is used, it means that each "node," that is, each computer on the network, carries the database. The network type is a peer-to-peer network [14].

If this ledger were on one server, it can be hacked more easily. Even if it is in a cloud, it could be hacked as the servers in a cloud all mirror each other. A distributed ledger is much resilient, as it not administered by a central server or entity.

If a transaction takes place and should be added to the blockchain, it needs to be verified. If everybody just added transactions, fraudulent or corrupt data could enter the blockchain. Therefore, a consensus mechanism is required [15]. If a node wants to add a transaction, it asks the other nodes for confirmation. The other nodes will vote for the update, using an algorithm. If more than a threshold, for example, 50%, of the replies are confirmations, the transaction will be added to the block.

3.4 Transparency and immutability

In blockchain technology, the network used is a peer-to-peer network. This means that each node is connected to the other node directly, not *via* servers as in a local area network (LAN). If the majority of these computers on the network have to agree on a change, a hacker would have to hack the majority of the computers in order to change one block and the subsequent blocks. While this is not absolutely impossible, it is considered not feasible in the computing industry [16]. This means in practice that the data in the blockchain cannot be changed. This is called immutability.

Furthermore, since a copy of the whole ledger is on every computer connected to the network, it is transparent.

3.5 Smart contracts

The "data" in the block can contain any kind of data. This includes copies of documents or software code. The software code may stipulate that when a certain event occurs, an action will be taken. For example, if the shipment is on board and the documents have been submitted to the blockchain and found to be correct, payment from the buyer to the seller should be made.

The oldest example of a smart contract is the vending machine: insert coins, choose a product, the product will be disbursed without human intervention.

4. How blockchain technology can help solve the problems

The problems identified with traditional supply chain at the beginning of this presentation were as follows:

- The origin of the raw material or the product is not obvious.
- Trust issues can exist between seller and buyer.
- The supply chain execution is inefficient.

When using blockchain technology in supply chain and logistics, each record in our blockchain contains a transaction and other data. This could be the bill of lading (B/L)—an electronic B/L or a scanned copy of a paper B/L—or just the fact that the container has been loaded onto a vessel. Each time a transaction happens, a record is added to the block.

This chapter will show how blockchain technology can help solving these problems.

4.1 The origin of the raw material or the product is not obvious

This problem can be solved easily by tagging the raw material at origin. A real-life example can be found in a case study published by Provenance [17] in which they describe the supply chain from tuna fishing to cans of tuna in a supermarket and a tuna dish offered in a restaurant. The blockchain will be created once the fish is caught on still on the boat, and at every step in the supply chain—unloading from boat to truck at the port, delivery to the processing plant, processing (which requires integration with the plants ERP to know which catch is packed into which cans), delivery and distribution, and finally arrival at the retail outlet or restaurant. The can of tuna shows a QR code which the consumer can scan with an app that connects to the blockchain. The supply chain will be followed backward to the fisher, and the consumer will receive a result on their smartphone confirming that the fisher acted in an ethical way.

Diamonds can be identified with 40 metadata points and a high definition of the diamond. The diamond can then be traced along the supply chain [18, 19]. This way, the buyer can ensure that their wedding ring does not contain a conflict diamond.

In order for the blockchain to confirm to the consumer that the product came from an ethical source, this information needs to be entered into the dataset. In the case of the tuna fish, the fisher needs to be approved by an independent agency, for example, an NGO. In the case of the diamond, the mining company must be approved. The parameters can be set as desired: Does this market segment require the produce to originate at an organic farm, or do they want produce that are not genetically modified, or do they not want any growth hormones in their meat? The origin can be certified by a trusted body, and the consumer can verify that the origin satisfies their requirements.

For industrial products, the origin can also be of interest, not only for the consumers. A car manufacturer can ensure that the components are genuine, for example [20, 21].

Customs departments require the presentation of a Certificate of Origin (C/O) in order to apply special privileges, such as import duty reductions or exemptions, to importers of goods made in countries with which a Free Trade Agreement exists. These C/O's are traditionally paper documents that have been signed and stamped by the exporter and the exporting country's issuing authority, often the Foreign Trade Department of the Ministry of Commerce. These paper documents can be forged.

If the issuing authority confirms on the blockchain that this shipment conforms to the rule of origin, the customs department of the importing country can easily see this on the blockchain. Also, the customs department in the exporting country can add the customs entry to the blockchain so that the customs department of the importing country can verify the declared product and the value [22]. Due to transparency and immutability, the origin cannot be changed.

4.2 Trust issues can exist between seller and buyer

The buyer does not need to believe that the products or components originate in a certain country; they can easily check it on the blockchain.

Once the blockchain has been created, the buyer can request access to the blockchain. They can then track the goods. Are they delayed in production or on time? Has the container been loaded on board the vessel, and has it departed on time?

A smart contract can be included in the blockchain so that when the goods are in accordance with the order, they have been loaded into a container, which has been transported to the port, and the container has been loaded onto the ship, and payment is triggered by the bank—which will also be on the blockchain. Trust is not required.

4.3 The supply chain execution is inefficient

It has already been suggested that certain conditions, when met, can trigger payments. While automated payments are already reality without blockchain technology—the telephone bill is an example—the difference here is the number of entities involved and the number of conditions that have to be fulfilled across the entities. Blockchain technology allows a higher degree of complexity.

Combined with artificial intelligence (AI) that manages demand forecasting, sales and operations planning systems and master productions schedules (MPSs) can be updated along with the material requirement planning (MRP). While this is within the ERP system and a blockchain is not required, the purchase orders (POs) that are sent to the suppliers leave the company's computing environment. POs can be sent by electronic data interchange (EDI) and then trigger an update of the supplier's MPS and hence his/her MRP. The supplier's ERP can then send their PO to their supplier (i.e., the manufacturer's second-tier supplier), and onward, upstream in the supply chain.

This requires different systems to cooperate. There are many ERP systems in the market, requiring different interface to even connect. In addition, the ERP system of any member of the supply chain may not be fully integrated internally [23] so that gaps in data transfer can occur.

There will be no visibility for the manufacturer into the second- or even third-tier supplier. This goes beyond the mere knowledge of the origin of the components and their origin in turn; visibility and data flow are essential to prevent the Bullwhip Effect [24].

Blockchain technology will also be able to take advantage of the Internet of Things (IoT) [25]. When the manufacturer has the forecast that was generated by AI and the new MPS has been set, manufacturing can be automated. The components are in the automated warehouse and can be retrieved and moved to the assembly line; assembly can be done with robots.

As the orders from the customers will be in the system, the system can generate the invoices and packing list and also submit the documents to customs. Customs can check for the business registration with the appropriate government office, and if any export licenses are required, this can be checked accordingly.

Once the goods are loaded into the container on the truck for delivery by seafreight, the driverless truck can be programmed to a destination inside the port. Once the truck arrives there—can be verified with smart CCTV cameras or satellite tracking—an automated crane can unload the container from the truck and store it in the container yard. The container yard storage system will be on the blockchain, along with the shipping line that has the confirmed booking (and the container will be moved alongside ship for loading) as well as customs.

During the voyage, the vessel data are available on the blockchain *via* satellite data. At the port of discharge, the port authorities, customs, and all other public and private entities concerned with this shipment receive the data they require, as they are all connected to the blockchain for this shipment. Immutability of data means that they all have the same data, and there will be no discrepancies. Yuan and Yong proposed blockchain-based intelligent transport systems, including multimodal transport [26].

This will greatly improve efficiency and eliminate fraud [27].

5. Possible problems with blockchains

The data are on all computers on the blockchain. A trader who imports goods will not want his customer to know his supplier or how much his actual cost were. The solution is to use a permissioned blockchain rather than a public blockchain. In a public blockchain, all data are visible to any node in the network. However, there is a solution: a permissioned blockchain requires permission to see certain data. According to Wüst and Gervais [28], a permissioned blockchain is useful if there are multiple writers, it is not guaranteed that everybody is only all the time, all writers are known, not all writers are trusted, and public verifiability is required. If some of these conditions are not given, they suggest permissionless public blockchains or even not to use any blockchain.

However, if a permissioned blockchain does make sense due to the number of contributors and the complexity of modern blockchains, there are two issues to consider: 1.) the consensus mechanism used and 2.) data security and integrity.

There are different kinds of consensus mechanisms, Proof-of-Work (POW) and Proof-of-Stake (POS) being the most well-known ones. They are fundamentally different but come to the same question: Who can vote whether a block is added? And also, how many votes are needed as a minimum, and what percentage is the threshold to accept the new block?

Data security and integrity have to be addressed, as we earlier said that the whole blockchain—with all data—is on every node computer. It is encrypted, and in the permissioned blockchain, only those data are accessible to a user for which they have received permission. The security is only guaranteed as the login credentials are kept secure.

In the supply chain, the wholesaler would receive permission to view the country of origin and whether the manufacturer is confirmed to not use child labor, for example. This is the information they need. Customs will need information about the exporter and the importer and any licenses, but not about the exporter's supplier. With the new German Supply Chain Act [29], German authorities will want to know whether the supply chain conforms to the regulations with regard to ethical manufacturing. Therefore, it is crucial that only those entities on the blockchain that need certain data have access to it. Another issue is that blockchains are often outsourced to cloud computing services for various reasons. Limited trust in cloud computing can reduce the adoption of blockchain technology [30].

Different blockchains use different technologies and are often not compatible with each other. However, this problem is being solved by creating interfaces between the blockchains. This means that if a manufacturer opts for blockchain provider A and their supplier uses blockchain provider B, and both use different technologies, the data can ideally still be moved from one technology to the other without loss of data or functionality. However, this does not apply to all types of blockchains, and even if there is a "fork"—a splitting of a blockchain for example to update the technology—is applied, access to both sides of the fork may not be possible [31].

Using blockchain technology can be expensive [30]. Depending on the blockchain type, there are different charges that apply for every transaction.

6. Conclusion

Supply chains are becoming more complex and more difficult to manage. Problems exist because manual processes are still used, such as paper-based bills of lading, different computer systems within a company do not integrate, or different computer systems used by the different members of a supply chain may not integrate with each other. Blockchain technology can help solve problems of complex supply chains. Use of blockchain will increase efficiency, increase supply chain visibility, and eventually prevent mistakes and prevent fraud.

Author details

Thomas E. Fernandez
University of the Thai Chamber of Commerce, Bangkok, Thailand

*Address all correspondence to: thomas.utcc@gmx.net

References

[1] Delfis, B. GEODIS Unveils its 2017 Supply Chain Worldwide survey [Internet]. 2017. Available from: https://geodis.com/fr//sites/default/files/2018-08/20170511_%20pr%20geodis%20white%20paper%20eng2431288865035225216.pdf. [Accessed March 6, 2022]

[2] Darbanhosseiniamirkhiz M, Khairuzzaman Wan Ismail W. Advanced manufacturing technology adoption in SMEs: An integrative model. Journal of Technology Management & Innovation. 2021;7(4):2012. DOI: 10.4067/S0718-27242012000400009

[3] Li W, Chen J. Manufacturer's vertical integration strategies in a three-tier supply chain. Transport Research Part E: Logistics and Transportation Review. 2020;**135**(March):101884. DOI: 10.1016/j.tre.2020.101884

[4] Nakamoto S. Bitcoin: A Peer-to-Peer Electronic Cash System [Internet]. 2008. Available from: https://bitcoin.org/bitcoin.pdf. [Accessed March 6, 2022]

[5] Somapa S, Cools M, Dullaert W. Characterizing supply chain visibility – A literature review. The International Journal of Logistics Management. 2018;**29**(1):308-339. DOI: 10.1108/IJLM-06-2016-0150

[6] Basha MB et al. Consumers attitude towards organic food. Procedia Economics and Finance. 2015;**31**(2015):444-452. DOI: 10.1016/S2212-5671(15)01219-8

[7] Sharma N, Singhvi R. Consumers perception and behaviour towards organic food: A systematic review of literature. Journal of Pharmacognosy and Phytochemistry. 2018;7(2):2152-2155

[8] Gerrodette T. The Tuna-Dolphin Issue. 2nd ed. Encyclopedia of Marine Mammals. San Diego: Academic Press; 2009;**2009**:1192-1195. DOI: 10.1016/B978-0-12-373553-9.00272-8

[9] Zutschi A, Creed A, Sohal A. Child labour and supply chain: Profitability or (mis)management. European Business Review. 2009;**21**(1):42-63. DOI: 10.1108/09555340910925175

[10] Falls S. Picturing blood diamonds. Critical Arts. 2011;**25**(3):441-466. DOI: 10.1080/02560046.2011.615144

[11] Lafitte, P. Horsemeat Scandal Highlights Lack of Traceability [Internet]. 2013. Available from: https://www.france24.com/en/20130212-france-findus-beef-horse-meat-lasagna. [Accessed March 6, 2022].

[12] Hao Y, Xiao L. Risk analysis of letter of credit. International Journal of Business and Social Science. 2013;**4**(9):199-299

[13] Rosic A. What Is Hashing? [Step-by-Step Guide-Under Hood Of Blockchain] [online]. 2020. Available from: https://blockgeeks.com/guides/what-is-hashing/. [Accessed March 5, 2022]

[14] Mills D et al. Distributed ledger technology in payments, clearing, and settlement. In: Finance and Economics Discussion Series 2016-095. Washington: Board of Governors of the Federal Reserve System; 2016. DOI: 10.17016/FEDS.2016.095

[15] Srivastava R. Mathematical assessment of blocks acceptance in blockchain using Markov model. International Journal of Blockchains and Cryptocurrencies. 2019;**1**(1):42-53

[16] Duy PT, Hien DTT, Pham V-H. A survey on Blockchain-based applications for reforming data protection, privacy and security. Hội thảo lần thứ III: Một số vấn đề chọn lọc về an toàn an ninh thông tin – Đà Nẵng. Vietnam. 2018

[17] Provenance. From Shore to Plate: Tracking Tuna on the Blockchain [online]. 2022. Available from: https://www.provenance.org/tracking-tuna-on-the-blockchain#pilot-phase-1. [Accessed March 5, 2022]

[18] Gordon S. Diamonds and the Blockchain [Internet]. 2018. Available from: https://www.babson.edu/academics/executive-education/babson-insight/technology-and-operations/diamonds-and-the-blockchain/. [Accessed March 6, 2022]

[19] Choi T-M. Blockchain-technology-supported platforms for diamond authentication and certification in luxury supply chains. Transportation Research Part E: Logistics and Transportation Review. 2019;**128**:17-29. DOI: 10.1016/j.tre.2019.05.011

[20] Abeyratne SA, Monfared RP. Blockchain ready manufacturing supply chain using distributed ledger. International Journal of Research in Engineering and Technology. 2016;**05**(09):1-10

[21] Kern J, Wolff P. The digital Transformation of the Automotive Supply Chain – an Empirical Analysis with Evidence from Germany and China: Case Study Contribution to the OECD TIP Digital and Open Innovation Project [Internet]. 2019. Available from: https://www.researchgate.net/profile/Johannes-Kern/publication/348189852_The_digital_transformation_of_the_automotive_supply_chain_-an_empirical_analysis_with_evidence_from_Germany_and_China_Case_study_contribution_to_the_OECD_TIP_Digital_and_Open_Innovation_project/links/5ff323b1a6fdccdcb82e5856/The-digital-transformation-of-the-automotive-supply-chain-an-empirical-analysis-with-evidence-from-Germany-and-China-Case-study-contribution-to-the-OECD-TIP-Digital-and-Open-Innovation-project.pdf. [Accessed March 6, 2022]

[22] Jensen T, Hedman J, Henningsson S. How TradeLens delivers business value with Blockchain technology. MIS Quarterly Executive. 2019;**18**(4):221-243. DOI: 10.17705/2msqe.00018

[23] Mahmood F, Khan AZ, Bokhari RH. ERP issues and challenges: A research synthesis. Kybernetes. 2020;**49**(3):629-659. DOI: 10.1108/K-12-2018-0699

[24] Forrester JW. Industrial dynamics--a major breakthrough for decision makers. Harvard Business Review. 1958;**36**(4):37-66

[25] Kasten JE. Engineering and manufacturing on the Blockchain: A systematic review. IEEE Engineering Management Review. 2020;**48**(1):31-47, 1 Firstquarter, march 2020. DOI: 10.1109/EMR.2020.2964224

[26] Yuan y, Wang F-Y. Towards Blockchain-based intelligent transportation systems. In: 2016 IEEE 19th International Conference on Intelligent Transportation Systems (ITSC). Rio de Janeiro, Brazil: Windsor Oceanico Hotel; 2016. November 1-4, 2016

[27] Hackius N, Petersen M. Blockchain in logistics and supply chain: Trick or treat? Published in: Digitalisation in Supply Chain Management and Logistics, Wolfgang Kersten, Thorsten Blecker and Christian M. Ringle (Eds).Berlin: epubli GmbH; 2017.ISBN 9783745043280, October 2017

[28] Wüst K, Gervais A. Do you need a Blockchain? In: Crypto Valley Conference on Blockchain Technology. IEEE Computer Society, Crypto Valley; 20-22 June 2018. DOI: 10.1109/CVCBT.2018.00011

[29] Bundesgesetzblatt J. Teil I Nr. 46. Gesetz über die unternehmerischen Sorgfaltspflichten in Lieferketten. 2021:2959-2969

[30] Syed TA et al. A comparative analysis of Blockchain architecture and its applications: Problems and recommendations. IEEE Access. 2019;7:176838-176869. DOI: 10.1109/ACCESS.2019.2957660

[31] Golosova J, Romanovs A. The advantages and disadvantages of the Blockchain technology. In: 2018 IEEE 6th Workshop on Advances in Information, Electronic and Electrical Engineering (AIEEE). IEEE; 2018. pp. 1-6. DOI: 10.1109/AIEEE.2018.8592253

Chapter 2

Safety of Food throughout the Supply Chain

Fraidoon Karimi

Abstract

Today, in addition to food security, food safety is very important, because one of the prominent causes of increasing diseases is changing people's eating habits and lack of food safety along the food chain. Vegetables and fruits, in all stages of the food supply chain, are contaminated with a variety of bacteria, viruses, protozoa, and parasites. The prevalence of many diseases has been recorded in the past as a result of eating fruits and vegetables. The results of extensive research conducted in this field around the world were used in this chapter, and melon is discussed in detail as an example. Research findings show that pathogens that lodge in the skin of melon can enter the melon flesh when cut and infect it. Similarly, in hot climates, the bacteria Salmonella, Listeria, and *E. coli* multiply easily in melon flesh at low pH and soon reach pathogenic doses. Therefore, appropriate hygiene measures should be considered along the food chain. These hygiene measures are considered very important from the perspective that very small contamination by campylobacter, *Escherichia coli*, and norovirus, for example, can cause a major health problem. Highly vulnerable people include pregnant women, young children, the elderly, and the sick.

Keywords: safety, supply chain, human health, microbial hazards, chemical hazards

1. Introduction

Food security and food safety are two basic concepts of human nutrition and, especially recently, have played an important role in the food industry and trade worldwide. According to the Codex's definition [1] "Food safety is the absence or presence of small amounts of hazardous substances in food that do not threaten the health of the consumer; these include microbial, chemical, and physical factors that are often not seen, such as bacteria, viruses, or chemical residues [1].

Melons (*Cucumis melo* L.) and watermelons (*Citrullus lanatus* L.) are important members of the Cucurbitaceae family. The term musk melons include various cultivars of melon. Commercial melons are evaluated based on their various characteristics, such as shape (spherical or oblong), skin color, and shape of the flesh [2]. More than 100 million tons of melons and watermelons are produced worldwide, 80% of which are from Asia. The main producer, from 2009 to 2011, is China with 60% of world production. However, the exports of this country account for only 0.13% of the total exports of the world [3]. Major exporters of melons have been reported from Mexico,

Spain, the United States, and Brazil [3]. In Europe, melons are grown mainly in southern regions, such as Spain, Portugal, Italy, and Greece. In Europe, 3 million tons of watermelon and 2 million tons of different melon cultivars are produced every year [3]. Melons and watermelons are not produced in Germany [2]. Afghanistan is one of the main centers of melon production, which is currently producing 15 different varieties of melons, mainly in the northern and southern parts of the country. These varieties include Jane Tour, Zermati, Bori Kaleh, Asqalani, Sabz-Maghaz, Alishabi, Sefal-Sar, Marpost, Ghazikhani, Andalak, Qandak, Alla-Poqaq, Arkani, Arkani-Garmah, and Sard-e-Lelmi [4].

Fresh vegetables and fruits are contaminated with a variety of pathogens along the production-to-consumption chain, resulting in dangerous diseases in humans [4]. For example, melon is one of the fruits that is known among people as a fever and disease stimulator that some doctors even include this fruit in the list of patients to avoid (personal interview, August 20, 2018).

Pathogens identified in melon include Salmonella, Listeria monocytogenes, and Enterohemorrhagic Escherichia coli (E. coli) [5]. Campylobacter, E. coli, and Norovirus can cause serious diseases [5].

A report prepared for the Codex Committee in 2011 states that between 1950 and 2011, 85 cases of the melon disease were recorded [1]. Some cases of the melon disease have also been reported in Germany [2]. Melons, on the other hand, do not appear to contain substances that are harmful to health. Therefore, in this study, it was assumed that external factors cause diseases in consumers, these factors are transmitted from the skin to the inner parts of the fruit when the fruit is peeled and gets contaminated. To test this hypothesis, various relevant sources, including research reports and articles, have been studied and their key concepts have been included in this chapter. Due to production and supply conditions, melon skin is in contact with germs and parasites, especially in melons that have large and rough skin. On the other hand, disinfection of melon fruit does not include the stages of production, supply, and consumption, and therefore the fruit remains contaminated and causes health problems.

2. Methodology

This study has been done using various first-hand sources that include reports and research articles. The study materials were collected in the first step, and in terms of logical relationship with the hypothesis, were divided into three categories: (1) very relevant, (2) relevant, and (3) less relevant.

3. Findings and discussion

3.1 Microbiology and pathology

3.1.1 Bacteria salmonella

It is a gram-negative bacterium that is common in nature and can survive in it for a long time. This bacterium is found in all types of animals and is one of the most important causes of diarrhea in humans. Food, at any stage of the production-to-consumption chain, can be contaminated with this bacterium through indirect contact with human or animal feces. This pathogen can survive in or on food for several months. Even frost cannot kill it [2]. Compared to other bacteria, it can grow in a variety of conditions. This bacterium can multiply at a temperature of 7–45°C, and the higher the temperature, the more it multiplies. The ideal temperature for the multiplication of these bacteria is 37°C. Only at temperatures above 60°C does the bacterium begin to die. Multiplication of this bacterium is also possible at pH 4–5 [2].

Humans are infected with Salmonella bacteria by eating contaminated food. But the infection can also occur through direct contact with a sick animal or human. The pathogenic dose for an adult is 104–106 Salmonella bacteria [6]. Depending on the amount of fat in the food and the immune status, even doses of less than 100 Salmonella bacteria can cause disease. After infection and the average incubation time, the first symptoms of the disease are observed after 12–36 h [6]. This bacterium causes diarrhea with abdominal pain, nausea, vomiting, and fever. In some cases, especially in vulnerable people, this bacterium causes severe illness and even death [6]. According to the Robert Koch Institute in Germany (RKI), Salmonella infected 17 people by eating melon, of which eight were hospitalized [6]. Vulnerable people include children under the age of five whose immune system has not yet been developed, adults, and those who already have a disease and are under treatment [2].

3.1.2 Listeria monocytogenes

It is a gram-positive bacterium and is a very important representative of genes Listeria. This bacterium is very common in nature and is found in soil, surface water, sewage, plants, and the digestive system of animals. This bacterium can contaminate food along with the production chain to the consumption of food, for example, during milking, butchery, vegetable or fruit collection; or cause contamination during food processing. Foods that are eaten raw have a higher risk of infection because they do not undergo any germicidal process after processing [7]. Listeria lives and multiplies in high temperatures and pH, and in a salty environment, and is a bacterium that is resilient and adapts to a variety of environmental conditions. It can multiply at 4–45°C and at pH 4.3–9. In addition, it can multiply at a salt concentration of 13% [7]. The ideal temperature for the multiplication of this bacterium is 30–37°C in neutral and slightly humid conditions, where the salt concentration is 0.5% [8].

The bacterium is transmitted to humans primarily by eating contaminated food. The dose of trauma depends on the condition of the immunity and the health of the person. According to European Union (EC) laws, ready-to-eat food is safe if the bacterial dose per gram is less than 100 calories (CFU) [6]. A healthy person does not develop a serious illness after an indication, but either develops cold symptoms or has indigestion. Unless people have a weak immune system, they can develop severe illness after several weeks of incubating the bacteria in their bodies. The bacterium causes blood poisoning, encephalitis, or meningitis, and the mortality rate from this bacterium is low. In pregnant women, it can lead to failure or premature birth and can cause irreversible disorders in infants. In Germany, 337 cases of Listeria's disease have been reported [9]. In addition to infants, people older than 60, especially men, developed leukemia. In 2011, 147 people in the United States became infected with Listeria, and 33 of them died [10]. In this country, Listeria bacteria was found in many samples of melons. The US Food and Drug Administration (FDA) has concluded that the disease is present in the melon peel and enters the inside of the fruit through the skin and is then transmitted to humans through eating the flesh [10]. According to European Union (EC) regulations, the concentration of Listeria in prepared foods should not exceed 100 colonies per gram of food.

3.1.3 Enterohemorrhagic Escherichia coli

A gram-negative anaerobic bacterium, abbreviated as E. coli, is capable of producing the poison ciguatoxins that cause bloody diarrhea in humans. The bacterium E. coli O157: H7 causes many diseases worldwide. This bacterium is present in the digestive system of ruminants, such as cattle, sheep, goats, and deer, and is transmitted to the environment through animal feces. This bacterium was resistant to drought, frost, and acidity, which means that it can survive in soil, water, and feces for several months [2, 10].

Food is contaminated with this bacterium at any stage of the production-to-consumption chain through direct or indirect contact with the feces of an infected animal or human. This bacterium multiplies at a temperature of 8–45°C and increases with increasing temperature. The ideal temperature for the multiplication of these bacteria is 37°C. Observation of E. coli O157: H7 in 0.5, 1, and 1.5% of organic acid (HCl) indicates that this bacterium also tolerates acidic conditions [2, 10]. In one experiment, using synthetic gastric extract, E. coli O157: H7 was observed to survive at pH 1.5 and it was found that this bacterium is also resistant to strong acidity [2].

The bacterium can also be spread through direct contact or by eating contaminated food by humans. The infectious dose of this bacterium is very low, that is, 100 of them cause disease. The incubation period is 2–10 days and averages 3–4 days. This bacterium causes watery diarrhea. In some cases, it causes swelling of the intestines and causes seizures and stomach pain, producing bloody stools, and in some cases, fever. Except after the disease is in the early stages, it does not show any symptoms [10]. In young children, the bacterium causes uremic hemolytic syndrome (HUS). In some cases, kidney failure, anemia, and thrombocytopenia can occur. These severe symptoms occur in 5–10% of patients. In the acute stage, the mortality rate due to this disease factor reaches 2% [2].

3.2 Pathogenic contamination

3.2.1 Melon consumption

Melons, like many other fruits and vegetables, can be produced both in the greenhouse and outdoors. After collection, it is either packaged directly on the farm or cooled before packaging. In arid regions (such as Afghanistan) packaging on the farm is common [11]. Melons are either sold whole or sliced, and in large stores, they are sliced and wrapped in plastic, which is sometimes mixed with other fruits. In Germany, the cooling of melon slices is prohibited [11].

According to estimates by the European Food Safety Authority [12], about 7% of European citizens consume melons [12]. It has also been observed that melons are consumed by all age groups, including children and the elderly, due to their deliciousness [13].

3.2.2 Contamination of melon fruits

Melons are grown on the ground. They need warm, sunny weather and moist soil. Due to the size and weight of its fruits, the fruits are in contact with the ground and are easily contaminated with pathogens, either through contaminated soil or water [14]. Studies in Mexico and the United States show that irrigation is a major cause of contamination [14]. Another study, conducted in Texas, USA, found that irrigation water was contaminated with 9.4% Salmonella bacteria [15]. Insects are also carriers of pathogens [16]. A study by Lopez-Velasco et al. [16] found that Salmonella could also enter plants through roots; they concluded that the possibility of root infection was not high and skin contamination was the main cause.

The fruit part that is in contact with the ground is a special area for contamination with bacteria and fungi. To reduce this contamination, the ground is covered with plastic, or fruits are placed in plastic bags. Another way to reduce contamination is to roll (turn over) the fruit repeatedly. This is done by the hands and if the hands are not clean, it will cause more contamination [17].

Contamination also occurs during collection and transfer, for example, by infected people. After harvesting, the fruits are cooled by air or water, in both cases, there is a possibility of contamination and spread of contamination [18]. Clean melons are also contaminated, especially when washed in the basin [18]. In a study by Duffy et al. [19], which compared washed and unwashed melons, it was found that washed melons were contaminated with Salmonella, because contaminated surfaces, as well as contaminated fruits, caused contamination of other fruits. Similarly, the Salmonella bacterium was observed in samples taken from cold storage [20].

To keep the fruit cool during transfer, ice is placed around it, which is usually melted and this causes the contamination of clean fruits [19]. All stages of melon production require human labor. The germs survive on the skin of the melon [17]. In particular, coarse skin is a good place to protect germs from washing and disinfection [18]. The larger the skin of the melon, the more germs it contains (**Figure 1**) [2].

3.2.3 Presence of pathogens in melon

In one observation, out of a total of 366 cantaloupe melons, 16 were contaminated with pathogenic microbes. Salmonella was observed in eight specimens, and Shigella in four specimens, while E. coli was not observed in this specimen [10]. In

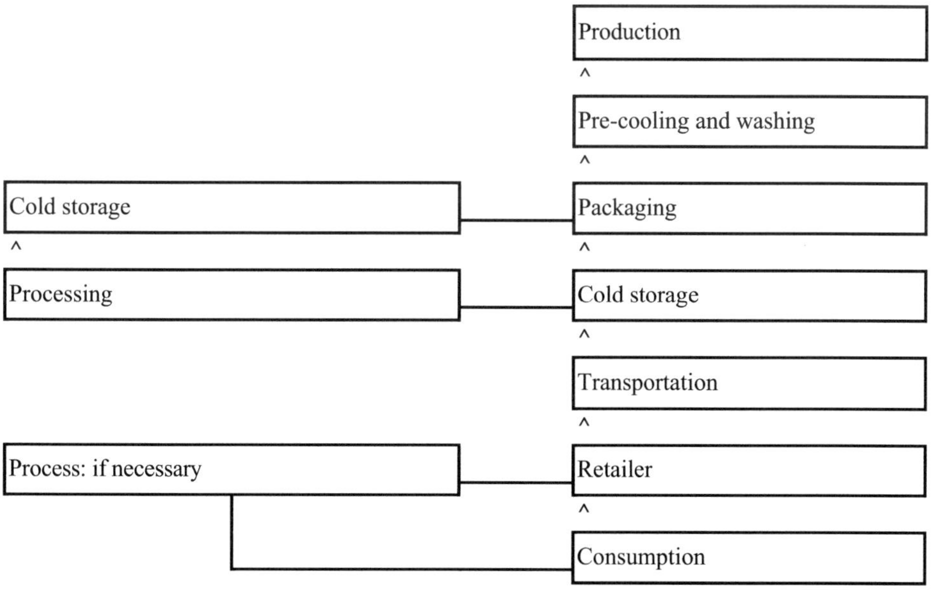

Figure 1.
Showing the food supply chain.

another study by the Austrian Institute of Technology [1], it was observed that the presence of Salmonella in melons is different [1]. In Mexico, 12 out of 55 samples were infected with Salmonella. In Nigeria, three out of every 50 melons were infected with Salmonella bacteria. In addition to Salmonella, other microbes, such as Staphylococcus aureus, Enterobacter aerogenes, and Klebsiella pneumoniae, were also observed [4]. In 2000, Castillo et al. [5] collected 950 samples of melons from Texas and 300 samples from Mexico; 0.5% of the samples were from Texas and 0.3% from Mexico were contaminated with Salmonella [10], respectively. Salmonella was also found in watermelons imported from Costa Rica and Brazil. In Germany, out of 406 samples, three melons were contaminated with Listeria.

3.3. Durability and growth of pathogens in the skin and flesh of melon

Pathogenic bacteria in melon skin survive for a long time and multiply [2]. The number of E. coli O157: H7 bacteria increases by more than 100 times in one square centimeter of melon skin and a temperature of 25°C in 4 days [4]. The large, rough skin of a melon provides a multiplication space for germs. Salmonella formed a biofilm in the skin of the cantaloupe melon at 20°C after 2 h, and the bacterium increased by 200% at room temperature for 24 h [2]. **Table 1** shows the amount and timing of the multiplication of Listeria and Salmonella bacteria in melons and watermelons at different temperatures. It is observed that the proliferation of bacteria has increased in proportion to the temperature; by increasing the temperature from 10 to 30°C, the largest number of these bacteria is produced in the shortest time.

Ukuku et al. [19], in their study, found that the number of aerobic bacteria in the flesh of the cantaloupe was higher than the number of these bacteria in the flesh of the honeydew melon. The microbes in the skin are transferred to the melon flesh during slice cutting [19, 21]. Due to the low acidity of the melon flesh, the bacteria survive and multiply rapidly. Listeria and Salmonella bacteria started multiplication at room temperature after 2 h [19], and the researchers also reported that, as a general rule, the higher the storage temperature, the shorter the bacterial division.

Studies by Li et al. [12] show that Salmonella and E. coli O157: H7 did not grow in the skin of cantaloupe, honeydew, and watermelon at 4°C, but at 15°C and higher. The proliferation of these bacteria began rapidly. E. coli O157: H7 began to multiply more

Pathogen	Temperature (°C)	Watermelon		Cantaloupe	
		Lag phase (h)	Generation time (h)	Lag phase (h)	Generation time (h)
Listeria monocytogenes	10	24a	13.03a	24a	7.12a
	20	18b	2.17b	6b	1.72b
	30	4c	1c	4b	0.84b
Salmonella enteritidis	10	24a	7.47a	24a	7.31a
	20	None	1.6b	none	1.69b
	30	2c	0.51c	2c	0.69c

Table 1.
Lag phases and generation times of Listeria and Salmonella bacteria in watermelon and melon [4].

rapidly than Salmonella at room temperature (23–25°C). Contrary to expectations, despite the different pH of watermelon (pH 5.1–5.6) and melon (pH 6.1–6.6), the same proliferation and growth of bacteria were observed in these two fruits. However, the growth rate of Salmonella and E. coli in honeydew melon was slightly lower than in other types of melons [19].

If the peeled melon is stored for a long time without refrigeration, the risk of germs growing in it increases [19]. Storing melon at 22°C caused severe growth of Salmonella. The number of Salmonella reached 1.3 cfu/g bacteria after 5 h and 0.2 cfu/g in a watermelon after 5 h [19]. In melons that were not refrigerated, the number of Listeria increased at 20°C compared to those refrigerated at 5°C. The researchers concluded that storing melon slices under normal room conditions, even for a limited number of hours, could increase Listeria bacteria [21].

According to scientific studies, washing with water does not significantly reduce the number of germs in melon skin [19]. Washing the cantaloupe melon for 3 minutes did not reduce the number of Salmonella and E. coli [2]. The researchers attributed the large surface area of the melon to the absorption, preservation, and contamination of the melon. To clean the surface of the melon, it is a good way to use a brush. A study by Parnell et al. [15] showed that brushing melon skin for 60 s reduced the number of Salmonella quadruples, and large-skinned melons reduced the number of bacteria by 1.6 times. After the melon was disinfected with 2.5% hydrogen peroxide, the number of microbes was reduced in it, except that Listeria bacteria is still observed in the melon flesh after spraying [19].

4. Conclusion and recommendations

Fruits and vegetables have caused diseases in the most developed countries. In the past, many cases of disease outbreaks have been reported after eating melons, some of which were very serious. For example, 147 people became infected with L. monocytogenes in the United States in 2011 after eating melons, including 33 deaths and a pregnant woman who was unable to give birth as a result of the disease. To reduce diseases, the following orders should be considered along the production chain to consume vegetables, fruits, and especially melon fruits.

During production, use plastic and/or organic mulch to reduce contamination, or place fruits in plastic bags. Another way to reduce contamination is to roll over fruits consistently. This is done by the hands and the hands should be clean, otherwise, it will cause more contamination [10, 21].

Studies show that peeling melon before eating significantly reduces germs [21]. In the United States, Canada, Australia, and New Zealand, the melon is cooled immediately after slicing as a precaution, and discarding melon slices that have been left in normal air for more than 2 h has been ordered [20, 22]. Therefore, observing these measures is recommended.

It is recommended that restaurants, wholesalers, and families cut and consume melon little by little, and melon slices should be consumed in less than 2 h. It is recommended that pregnant women and other vulnerable people, such as young children, adults, and the sick, refrain from eating melon that has been stored outdoors for several hours.

Hygiene measures should be taken when cutting the melon, such as slicing should be done in a wider and clean place with clean hands, and a clean knife and a clean container should be used to prevent contamination of the melon flesh. It is very important to consider home hygiene and personal hygiene because the infection dose of Listeria bacterium is very low (100 colonies). In addition to maintaining good hygiene, it prevents the melon from becoming contaminated with other germs, such as campylobacter, type A jaundice, and norovirus.

To prevent contamination and disease, retailers, restaurants, cafeterias, and others should cut a small amount of melon and consume it in less than 2 h.

It is recommended that the melon husk be eaten immediately or put in the refrigerator immediately. The suitable temperature for the refrigeration of melon cut is 4-6 Degree Celsius as it minimizes the multiplication of germs. Pregnant women and people with weak immune systems, such as children, the elderly, and the sick, should refrain from eating peeled melons that have been left in the room for more than 2 h.

In the field of exports, since the issue of food safety is more important today than food security, producers and exporters must consider hygiene throughout the stages of the supply chain, because competition and continuity in international markets are possible only if the fruits are healthy and free of pathogens and dangerous elements, such as Argentum, lead, and arsenic; also, free of residues of agricultural chemicals. Therefore, exported fruits must comply with the standards of the global quality control organizations, especially the food standards of the Codex Alimentarius Commission, and HACCP.

Author details

Fraidoon Karimi
Faculty of Agriculture, Department of Horticulture, Al Beroni University, Afghanistan

*Address all correspondence to: frdmfrdn@gmail.com

References

[1] FAO/WHO. Microbiological hazards and melons. In: Report Prepared for: Codex Committee on Food Hygiene Working Group on the Development of an Annex on Melons for the Code of Hygienic Practice for Fresh Fruit and Vegetables (CAC/RCP 53-2003). 2011

[2] BfR. Melons: Health Hazard through Contamination with Pathogenic Bacteria. BfR Opinion 021/2013. 2013. Available from: http://www.bfr.bund.de/

[3] FAOSTAT. Large time-series and cross-sectional data relating to hunger, food and agriculture for 245 countries and territories and 35 regional areas, from 1961 to the most recent year. Production and Trade. 2013. Available from: http://faostat3.fao.org

[4] Karimi F, Aria F, Qarluq AQ. Studies on Food Safety of Melon in its Food Supply Chain. International Journal of Innovative Research and Scientific Studies. 2019;2(3):48-57. Available at SSRN: https://ssrn.com/abstract=3474214

[5] Castillo A, Mercado I, Lucia LM, Martínez-Ruiz Y, Ponce De León J, Murano EA, et al. *Salmonella* contamination during production of cantaloupe: A binational study. Journal of Food Protection. 2004;**67**:713-720

[6] Arnold KW, Kaspar CW. Starvation- and stationary-phase-induced acid tolerance in *Escherichia coli O157:H7*. Applied and Environmental Microbiology. 1995;**61**:2037-2039

[7] Khelef N, Lecuit M, Buchrieser C, Cabanes D, Dussurget O, Cossart P. *Listeria monocytogenes* and the Genus *Listeria*. In: Dworkin M, Falkow S, Rosenberg E, Schleifer KH, Stackebrandt E, editors. The Prokaryotes: Volume 4: Bacteria: Firmicutes, Cyanobacteria. New York: Springer; 2006. pp. 404-476

[8] Shabala L, Lee SH, Cannesson P, Ross T. Acid and NaCl limits to growth of *Listeria monocytogenes* and influence of sequence of inimical acid and NaCl levels on inactivation kinetics. Journal of Food Protection. 2008;**71**:1169-1177

[9] McClure PJ, Kelly TM, Roberts TA. The effects of temperature, pH, sodium chloride and sodium nitrite on the growth of *Listeria monocytogenes*. Journal of Food Microbiology. 1991;**14**:77-91

[10] RKI, editor. Infektion sepidemiologisches Jahrbuch meldepflichtiger Krankheiten für 2011. Berlin: 2012. Available from: <www.rki. de > Infektionsschutz > Infektionsepi. Jahrbuch > Jahrbücher>

[11] FDA. Factors potentially contributing to the contamination of fresh, whole cantaloupe implicated in the multi-state *Listeria monocytogenes* . Foodborne Illness Outbreak. FDA. 2011. pp. 1-17

[12] Li D. Development and Validation of a Mathematical Model for Growth of *Salmonella* in Cantaloupe. Canada: Rutgers University, Graduate School-New Brunswick; 2012. p. 20

[13] Panel on Biological Hazards (BIOHAZ). Scientific opinion on the risk posed by pathogens in food of non-animal origin. Part 1 (outbreak data analysis and risk ranking of food/pathogen combinations). EFSA. 2013;**11**:3025

[14] Richards GM, Beuchat LR. Attachment of *Salmonella poona* to cantaloupe rind and stem scar tissues

as affected by temperature of fruit and inoculum. Journal of Food Protection. 2004;**67**:1359-1364

[15] Parnell TL, Harris LJ, Suslow TV. Reducing *Salmonella* on cantaloupes and honeydew melons using wash practices applicable to postharvest handling, foodservice, and consumer preparation. Journal of Food Microbiology. 2005;**99**:59-70

[16] Velasco L, Sbodio G, Tomas-Callejas A, Wei A, HupTan P, Suslow K. Assessment of root uptake and systemic vine-transport of *Salmonella enterica* sv. Typhimurium by melon (*Cucumis melo*) during field production. Journal of Food Microbiology. 2012;**158**:65-72

[17] Anonymous. Food of plant origin: Production methods and microbiological hazards linked to food-borne disease. Lot 1: Food of plant origin with high water content such as fruits, vegetables, juices and herbs. CFT/EFSA/BIOHAZ/2012/01. Supporting Publications; 2013. EN-402

[18] Ukuku DO, Olanya M, Geveke DJ, Sommers CH. Effect of native microflora, waiting period, and storage temperature on Listeria monocytogenes serovars transferred from cantaloupe rind to fresh-cut pieces during preparation. Journal of Food Protection. 2012;**75**:1912-1919

[19] Duffy EA, Lucia LM, Kells JM, Castillo A, Pillai SD, Acuff GR. Concentrations of *Escherichia coli* and genetic diversity and antibiotic resistance profiling of *Salmonella* isolated from irrigation water, packing shed equipment, and fresh produce in Texas. Journal of Food Protection. 2005;**68**:70-79

[20] Annous BA, Burke A, Sites JE, Phillips JG. Commercial thermal process for inactivating *Salmonella Poona* on surfaces of whole fresh cantaloupes. Journal of Food Protection. 2013;**76**:420-428

[21] Vadlamudi S, Taylor TM, Blankenburg C, Castillo A. Effect of chemical sanitizers on *Salmonella enterica serovar poona* on the surface of cantaloupe and pathogen contamination of internal tissues as a function of cutting procedure. Journal of Food Protection. 2012;**75**:1766-1773

[22] Health Canada. Safe handling of fresh fruits and vegetables. Report. 2009. p. 18

Section 2

Logistics

Chapter 3

Harnessing the Environment of Maritime Transport and Port Logistics Sector in the Management of Covid-19 Pandemic

Theophilus Chinonyerem Nwokedi

Abstract

The outbreak of Covid-19 pandemic led to huge loss of human and material Capital and adversely affected the economies of nations. The fight against the spread has been serious and strategically coordinated and implemented across major Countries; but exposure to its spread has continued to raise fears about the possibility of achieving total eradication. Public health experts opine that the disease may not totally go away soon, and as such, society should learn to live with it as the new normal. The study thus evaluated the influence of harnessing the built environment of maritime transport and logistics sector in the management of exposure to spread of the disease. It identified the various components of the built environment types in the maritime transport sub-sector that represent potential nodes of maritime workers exposure to the spread of Covid-19 as well as the Covid-19 infective pathways in the maritime sub-sector and analyzed the risk of exposure each infective pathway and node poses to the exposure to Covid-19 spread. It also identified and analyzed the risk management approaches to be employed in limiting exposure to the disease in the maritime and port logistics sector in Nigeria.

Keywords: built-environment, maritime-transport, port-logistics, spread, Covid-19

1. Introduction

1.1 Overview of the role of built environment of maritime transport and port sector in management of Covid-19 pandemic

The environment is defined as the ecosystem, habitat and/or living place of any organism. It is the totality of the surroundings and all the content of the surroundings in which an organism lives including the natural forces and other species of livings things that create interdependency relationships for the organism's development and growth. The environments of an organism thus have or should be made to have inherent capacity to shield and protect it from the danger of injury, health-hazard, damage and/or death from unnatural causes. Man over the ages have artificially harnessed and made habitable for himself environments that were initially considered unfavorable for human habitation by the use of technology. One such example is the marine environment, sea and/or offshore locations which were natural habitats for aquatic organism and other marine species. The exploration and exploitation of the resources of the marine environment over the years has led to the development of ships and marine structures of many kinds that support human adaptation and living in the marine environment in the course of his occupation. The same can be said for the urban and city centers of today which were initially natural forests turned built cities and urban centers by the entrepreneurial activities of humans. Thus we view the concept of the built environment as a concept that encompasses artificially made structures, platforms, buildings; urban, suburban and rural settlements and the relative facilities that accommodate humans to live, work and carry-out diverse socio-economic, political and all forms of human activities that are necessary for and support collective human existence [1]. It therefore behooves on human operators to ensure that the built environment have capacity to maximize the protection of humans inhabitants from the danger of injury, health hazards, damage, and death from unnatural causes. This may be achieved by employing various means and strategies, but mostly the planned management approach which ensures that outcomes remain in line with behavioral objectives. According to Roof and Oleru [2], the concept of built environment is not limited to urban and suburban housing settlements that provide shelter for times spent indoor, it equally encompasses shelter for times spent at work related environment as more than 5% of average North American's workers time is spend in car.

In the Nigeria maritime industry for example, navigators and seamen live in ships and offshore structures for as long as 3 months before the next change of crew while dock workers and terminal operators spend consecutive 8 h work period each day in the seaports and terminal infrastructures in Nigeria. Thus ships, the seaport infrastructures, the terminals, the offshore structures etc. form the built environment of the maritime transport and ports industry in Nigeria whose capacity to provide protection to the inhabitants and users, cum shore-based stakeholders and/or contribute to programmes and schemes aimed at eliminating and curtailing the spread of life threatening infections, like the current Covid-19 pandemic must be enhanced.

The environment of maritime transport and port logistics sector is therefore, viewed as the totality of the maritime ecosystem including the sea, the coastal water zones, the inland water transport (IWT) zones, upon and/or in which the cargo ships, offshore floating and fixed productions, storage and offloading (FPSO) systems, fishing vessels, cruise ships, barges, river crafts, dredgers, seaports and Inland River ports, marine terminals, shipyards and docks, etc. as built maritime transport and ocean exploitation support structures, that accommodate maritime workers; that harness the ships and watercrafts for waterborne transport and other related operations. For purposes of developing models to enable the maritime sub-sector successful manage the fight against the spread of the Covid-19 pandemic; we summarized the built environment of maritime transport system that must be involved in the successful management of the exposure to and spread of the Covid-19 pandemic as shown in **Figure 1**.

Figure 1 presents the built environment of the maritime transport system, indicating the various ships, port infrastructure and shore based maritime structures accommodating human activity types in the maritime industry while also interacting with, affecting and impacting on activities, operations, process and life in the various

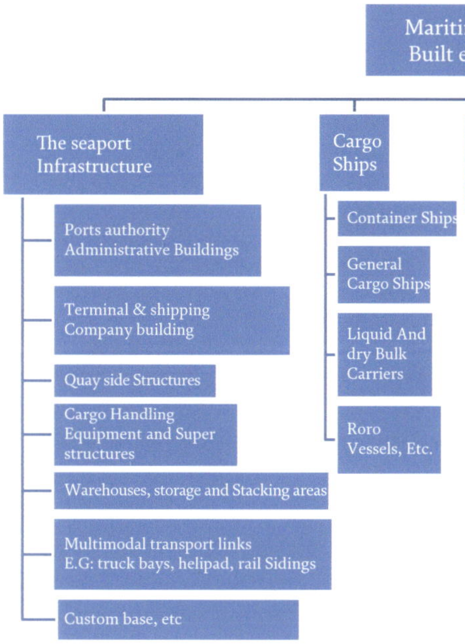

Figure 1.
Summary of the built environment of maritime transport showing the hierarchy of relationships. Source: Prepared by Nwokedi [3].

urban, suburban and rural land based settlements. They protection of the inhabitants of the above environment of maritime transport and the urban, suburban and rural land based settlements is the motivation for harnessing the maritime and port logistics industry against the spread of the Covid-19 pandemic [4]. Any management model or strategy that did not holistically involve the identified components of the environment of maritime transport, may not succeed, as the un-captured/uninvolved component or sub-system may end up re-infecting the entire maritime transport system, and subsequently, the land based urban and suburban settlements.

The China Country office of the World Health Organization [5] in Wuhan City was the first to identify in humans and report a novel type of corona virus disease in December 2019 which was afterwards officially referred to as the Covid-19. The Covid-19 disease spread geometrically with large numbers of confirmed cases in many parts of World within a very short time causing it to be characterized as a global pandemic. Since the Covid-19 disease is currently determined not to be an airborne disease, humans are determined to be the commonest direct agent of its spread among human populations and objects in the built environments and human settlements. By implication, transports infrastructures (maritime transport, air transport, road, rail, etc.) remains the major means of the transmission of the disease across international borders and among local populations. Consequently, the built environment of the maritime transport and shipping sector like other transport modes constitute potential and real major agents of spread of the virus; following which entire built environments of maritime transport must develop proactive instruments and strategies for managing the environment as identified above to curtail the spread the spread of the Covid-19 pandemic. Nallon [6] notes the importance of quarantining ships for 14 days before enter the destination port following the first confirmed case of coid-19 disease onboard a container ship Gjertrud Maersk. The report notes that quarantining vessels will impact global seaborne trade negatively, that however is in line with management guidelines needed to ensure most importantly that the transmission and spread of the disease through the entire maritime industry to land based settlements environments is prevented and/or curtailed. Developing Covid-19 transmission and infection likelihood/risk model based on empirical evidences of the spread of the disease for seaports in Nigeria and ships calling to the ports as well as a framework and action plan for the entire maritime transport environment for the management and prevention of the spread of Covid-19 in Nigeria constitute the central aim of the study.

1.2 Covid-19 transmission modes and implications for management of the environment of maritime transport and ports

The Nigerian center for infectious disease control [7, 8] identified two major modes of transmission of Covid-19 disease to include:

i. Direct transmission: This involves human-to-human direct transmission which results when infected person comes in contact within about 6 feet with other people, respiratory droplets from the nose produced by sneezing, droplets from the mouth during speech, may drop on objects or be inhaled into the lungs by unprotected close-by persons leading to infection [7, 8]. The Bangladesh Maritime Authority gave a pictograph summary of direct human-human transmission as shown in **Figure 2**.

ii. Indirect Transmission: This involves objects-to-human transmission. Indirect transmission or object-to-human transmission occurs when humans come in contact with infected objects. Infected objects in this case are objects contaminated by droplets from noses and mouth of infected individuals in the course of sneezing and talking (**Figure 3**).

The implication of this in the management of Covid-19 in the maritime environment is that the mode specific transmission mean/routes must be identified and

Figure 2.
Direct (person-to-person) transmission of Covid-19. Source: Bangladesh Maritime Authority [9].

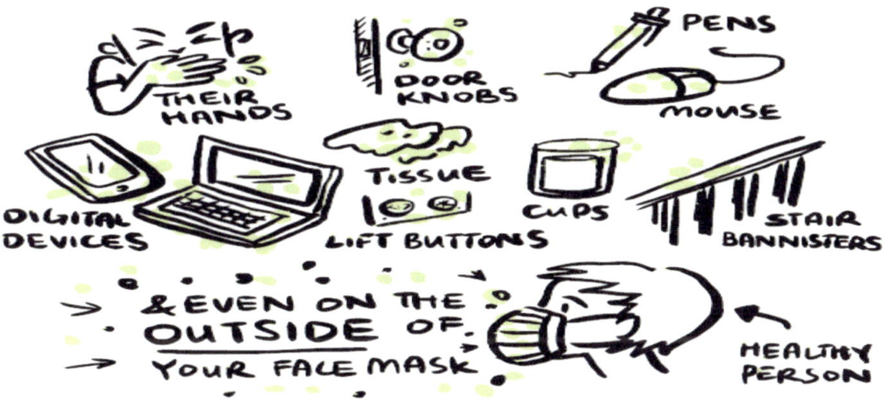

Figure 3.
Indirect (object-to-person) transmission of Covid-19. Source: Adapted from Bangladesh Maritime Authority [9].

barriers/shields built against it these transmission means and routes of transmission. The transmission barriers and/or shields takes the forms of regulatory instruments aimed at regulating the behavior of operators and manners of operations and use of equipment to yield expected outcomes. The expected outcomes being to break the transmission curve and achieve a Covid-19 free maritime environment and society. Implementing the use of personal protective equipment (PPE) and developing policies that limit risky behavior is also important. While the practice of 14 days quarantining cum isolation and treated of confirm cases in line WHO [5] guidelines will apply to human agents (seafarers and maritime workers), massive disinfection of non-human agents (cargo, ship surfaces, superstructures, equipment, etc.) is applied in management and treatment of such non-human agent. But since it is impossible to determine which non-human agents are already contaminated with the virus, risk assessment based on ship exposure to ports in regions with confirmed cases becomes necessary such that all such objects and ships identified as having visited a high risk region is mandatorily quarantined and disinfected before being allowed access to seaports. Based on the very nature of ship movements and the structure of the maritime industry operations, we identified Covid-19 transmission routes in the maritime industry for purposes of developing transmission barriers as a management strategy as follows;

i. External transmission routes: This is the transmission and infection of healthy ship and/or the occupant seafarers and objects onboard by parties external to the ship at any given point in time. It equally denotes a situation where an infected ships or crew transmits and infects healthy ports, settlements and other healthy ships with which it had contact. See **Figure 4** for a pictograph example of external route/channel Covid-19 transmission routes in the maritime industry.

   Major types of external Covid-19 transmission routes in the maritime industry include:

   1. Ship-to-shore (port)-transmission (inclusive of direct or indirect transmission)

   2. Port (shore)-ship-transmission (inclusive of direct or indirect transmission)

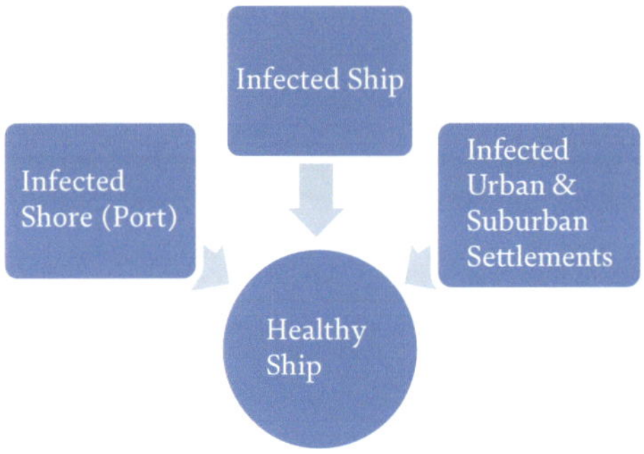

Figure 4.
Pictograph showing external Covid-19 transmission routes/channels in the maritime industry. Source: Nwokedi [3]. Note: The direction of the arrows is reversed in the case that an infected ship is transmitting and infecting healthy ports, urban settlements and other healthy ships.

3. Ship-to-ship-transmission (inclusive of direct indirect transmission)

4. Direct and indirect Urban and suburban human settlements-port-transmission (city and sub-city settlements-to-port-transmission).

ii. Internal transmission routes: This is the internal transmission and infection of healthy persons and objects within the same vessel or ports by individuals who are already carriers of the viral disease. For example, within a given seaport, an infected dockworker in the customs base may spread and infect the entire seaport community as shown in **Figure** 1, within a custom base, an infected officer will transmit and infects healthy colleagues which may lead to a situation of community infection, etc. see **Figure** 5:

Major types of internal Covid-19 transmission routes in the maritime industry include:

1. Internal ship-based-person-to-person-transmission onboard,

2. Internal ship-based object-to-human-transmission (within the same ship)

3. Internal Port-based human-to-human-transmission within the seaport and all the shipping companies, trucking units, etc. within the seaport.

4. Internal shore-based object-to-human transmission.

Identifying clearly the transmission routes in the maritime industry enables route specific Covid-19 transmission and exposure risk analysis to be carried-out and the likelihood/probability of transmission of and exposure to Covid-19 determined for both external and internal routes. All routes may thus be presented in term of the hierarchy of the probability transmission and exposure risks to guide operations and operators in the industry. For indirect transmission, disinfection by use of approved sanitizing agents in line with WHO guideline can be carried prioritizing high risk routes. Quarantining seafarers and other industry personnel from high risk areas for 14 day and isolation of confirmed cased followed by treatment in line with WHO guidelines is adopted in managing direct human to human transmissions. As can be seen both the external and internal routes of Covid-19 transmission and infection

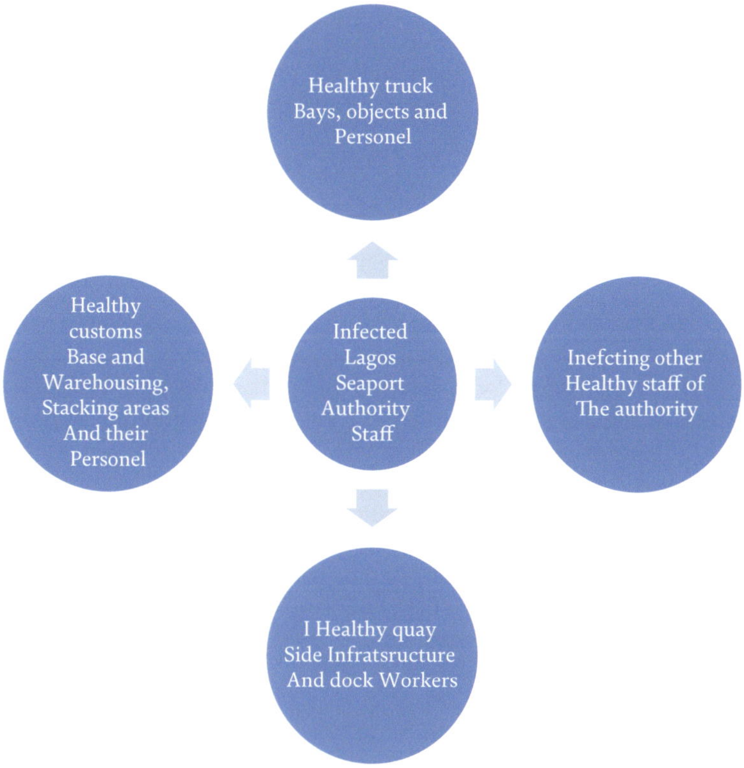

Figure 5.
Source: Nwokedi [3]. Note: The direction of the arrows indicating the direction of transmission is reversed in the case that an infected sub component of the seaport system is transmitting and infecting healthy port authority staff and other healthy sub-components of the port as identified in ***Figure 1****.*

spread in the maritime industry revolve around object transmission and human transmission. This suggests that among other things, high risk transmission and infection routes should prioritized in the serious implementation of the guidelines for the management of both direct and indirect transmission in the maritime transport sub-sector in Nigeria. The development of a framework for the management and control of the spread of the Covid-19 pandemic in the Nigerian environment of maritime transport will considered in the subsequent sections of the chapter.

The identified external and internal routes/channels of transmission among the built environment of the maritime transport and shipping sector further suggests a transmission-route specific approaches can be developed involving holistically all the components of the built environment of maritime transport including seaports, ships, shipyards, shipping companies and all the components earlier identified. This should commence with Covid-19 infection risk assessment in which a proximity based model may be used to assess risk of exposure of each port facility, ship, company facility, etc. to risk of Covid-19 infection. Higher level of caution and preventive strategies in line with WHO guidelines will thus be adopted by ships when calling to a port facility with higher risk of infection while ships with higher risk of infection determined by their proximity and duration and frequency of call to high risk ports are quarantined in line with available regulations as they call to healthy ports. To effectively curtail the transmission and spread of Covid-19 disease within the maritime sub-sector and between the maritime environment and residential settlements beyond the maritime environment, there is serious need for each component of the built environment in the maritime sector representing a node and/or point of transmission and infection development node specific strategies in line with general guidelines. This will equally be useful in determining the risk of exposure and infection at each specific node and other interacting nodes. See **Figure 6**:

Based on the above interaction model above, proactive management approach should be adopted where: At Node-A Seaport authority should:

i. Port Covid-19 exposure risk assessment (PCRA)

ii. Develop a port Covid-19 prevention plan (PCPP) encompassing action plans to be carried out by all agencies, companies, operators and stakeholders in the port including interaction principles and procedures with residential settlements (individual accessing the ports from residential settlements daily). The plan should be in line with WHO and NCDC general guidelines for control of the spread including but not limited to quarantining, isolation of confirmed cases, social distancing rules, use of face masks, etc.

iii. Appoint port Covid-19 officers and or prevention and management committee with the responsibility of ensuring the implementation of port Covid-19 prevention and management plan. Such officers must work in unison with the port health department.

iv. Determine which components or sections of the seaport pose high risk of transmission and infection of the entire seaport so that appropriate controls are deployed.

v. Determine what equipment is necessary for the prevention of the spread and provide all such equipment including PPE.

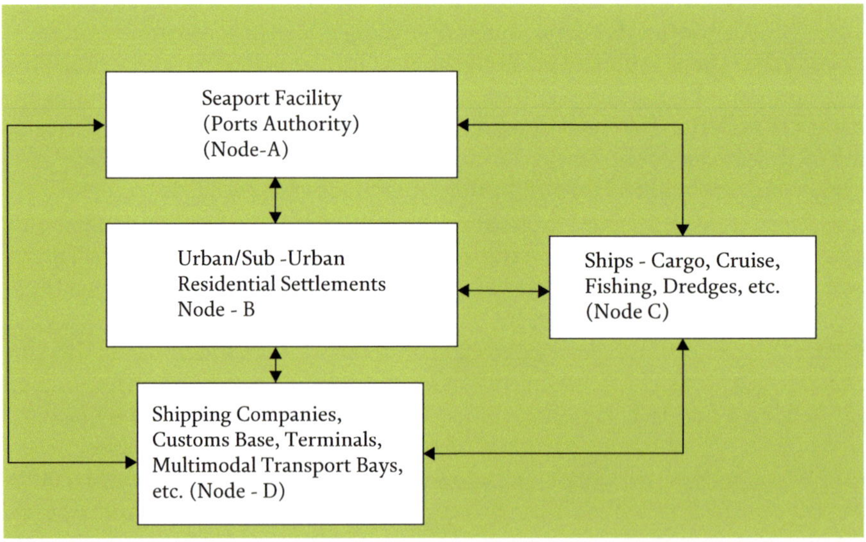

Figure 6.
Built environment of maritime transport and non-work residential settlements interaction model depicting Covid-19 transmission and infection nodes as it affects the maritime industry. Source: Nwokedi [3].

vi. Regulate the implementation of port Covid-19 prevention plan (PCPP) by all stakeholders and ships calling to the port.

vii. Carry-out performance appraisal routinely to ascertain the capacity of the current PCPP in curtailing the spread of the disease within the environment and improve if need be; among others functions.

At Node-B, Port users and other interest groups (e.g.: seafarers and their families, families of dockworkers and agencies personnel in ports, etc.) that interact with the built environment of the maritime transport from residential settlements represent potential transmission agents captured in node-B. To prevent transmission and infection from these groups requires that:

i. Port authorities, shipping companies, ships and other agencies in the maritime built environment as identified in that model of interaction develop access control programme for all such individual such that only those with serious business in the maritime environment are allowed access.

ii. Provide PPE and implement compulsory use of such by all accessing the facilities from residential settlements.

iii. Disinfect all objects entering the seaport facility and the component nodes of interaction from residential settlements.

iv. Develop policy blueprint on compulsory expected behavoiur pattern (CEBP) for all employees in the maritime sector accessing the built environment of maritime industry (seaports, shipping companies, customs base, truck bays, ships, etc.) from residential settlements in lines with World Health Organization guidelines.

At Node-C, Ships of various kinds calling to seaports and the ship-owners takes responsible of the fight against the spread of the Covid-19 disease. It is expected that each ship must develop strategic action plans aimed at not only avoiding being infected in course of interaction with ports, settlements and other maritime industry stakeholders, but to ensure that the ship, the crew and the objects on board does not transmit and infects healthy ports and residential settlements with the disease. To ensure this, each ship should:

i. Carry out ship exposure to Covid-19 risk assessment using proximity based model in which the ship assesses its closeness and/or contact with infected port regions.

ii. Develop a ship Covid-19 prevention plan (SCPP) in line with guidelines issued by the WHO and the local port of call including the 14 days quarantining, isolation and Covid-19 management principles.

iii. Develop schedule for routine disinfection of onboard objects and other ship surfaces following visits to ports in high risk regions.

iv. Appoint a ship-based Covid-19 prevention officer as the responsible authority for enforcing the SCPP.

v. Implement the port specific regulations for ships for the prevention of Covid-19 transmission and infection.

vi. Determine ship equipment requirement including PPE and acquire such.

vii. Determine rules for engagement of external parties to avoid the spread of the disease.

viii. Cary-out routine performance appraisal to evaluate how the SCPP was able to curtail the transmission to and infection of the ship and crew and recommend improvement.

At node-D, shipping companies, custom bases, multimodal transport bays, terminal operators and other stakeholders in the maritime industry have responsible to manage and prevent the spread of the Covid-19 disease by implementing the port authority regulations designed for each group of operators as well as company specific prevention rules, in line with the guidelines of the WHO. Each company and agency in this category should:

i. Carry-out company Covid-19 risk and exposure assessment (CCRA)

ii. Develop company specific Covid-19 prevention plan (CCPP)

iii. Appoint a company based Covid-19 prevention officer responsible for ensuring the implementation of the CCPP.

iv. Achieve the implementation of the PCPP as it affects the company.

v. Carry-out routine performance appraisal to determine the viability or otherwise of the CCPP; among other duties.

It is important to note the similarity between the approach to Covid-19 prevention in the maritime industry and the international ship and port facility security (ISPS) code implementation approach which outlines the roles of all interacting stakeholders in the security architecture of the maritime industry [10]. Government may thus for an apex organization enforcing and supervising the overall implementation of the Covid-19 prevention rules across the maritime industry through the Nigeria Maritime Administration and Safety Agency (NIMASA), the port health (PH) and National center for disease control (NCDC).

2. Relevance and importance of the environment of maritime transport and shipping IN management of the spread of Covid-19

The Nigeria center for disease control [7, 8] notes that the index case of Covid-19 disease in Nigeria was an Italian Citizen who arrived Lagos after being possibly infected the virus in his home Country, Italy. This evidences the roles of the overall transport industry inclusive of the aviation, road, water/maritime transportation, rail, etc. in the spread of the disease and thus formed the overall reason for the ban of international, regional and interstate travels as a combative measure against the spread of the Covid-19 pandemic. Since transport facilitates social political and economic interaction between and among diverse geographical locations, residential settlements and work/industrial zones, it has inherent potential to negate the social distancing rules if not deliberately strategized to ensure the observance of social distancing rules. As an industry with capacity to mobilize mass movement of people and goods across spatial locations, the build environment of the maritime transport sector is a major relevant stakeholder in the fight against the transmission and spread of the Covid-19 disease. As such, for the continued operation of this important sector, sector specific cum institution specific, policies, regulations, strategies and guidelines must be developed in line with the general guidelines and rules of the World Health Organization and the Nationals guidelines for the prevention of the spread of Covid-19. The maritime transport industry like the aviation and road transport subsectors, similar to the health institutions need to be viewed as front line sectors that are directly affected by the pandemic, directly involved in the spread of the disease from one region to another and to residential settlements. As such, the environment of transport (including maritime transport) is relevantly involved and must play frontline roles in the management of the transmission of Covid-19 in the society. As a core component of the work and/or industrial zone with interactive relations with diverse urban and sub-urban residential settlements, the built environment of maritime transport can significantly contribute in breaking the curve of transmission of the disease to the benefit of society. The relevance and importance of the environment of maritime transport in the fight against the spread of the Covid-19 pandemic is better understood by observing the overlapping nature of interaction between and among different forms of built environment consisting of the work/industrial environment, residential environment, social environment, and religious environment as shown in **Figure 7**. The tick blue areas indicate areas of intersections and union

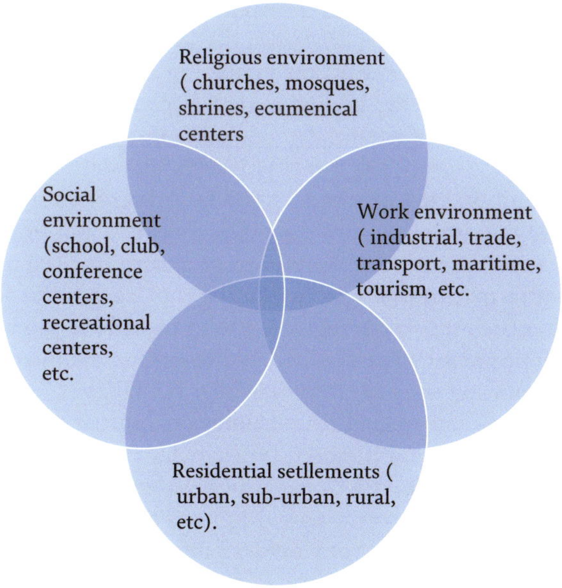

Figure 7.
Spatial interaction model depicting the relevance and importance of the environment of maritime transport and other spatial settlements in the management of Covid-19. Source: Nwokedi [3].

between and among the environments. Thus, it behooves on all environment types to be actively involved in the implementation of the measures aimed at preventing the spread of the virus. Otherwise, gain made in other environments by strict and active implementation of the guidelines may be eroded be interacting sections such that the overall built environment will be re-infected through one section of the system.

3. Harnessing the environment of maritime logistics sector in the fight against exposure to and spread of Covid-19: evidences from available literature

The novel Covid-19 pandemic is seen to be currently ravaging all aspects of life in global communities; major economic institutions, organizations, industries and sectors. Global and local trend social, political and economic interaction was altered in favor of online and/or virtual interaction following the introduction of social distancing and ban on international and local interstate and regional travels in a build to break the curve of transmission and infection across built environments. In the economic sector for example, major anchors of business development and growth as has been negatively impacted following the lockdown policies of many countries which caused output losses occasioned by production time losses by many businesses. The global transport, logistics and supply chain inclusive of the maritime and shipping logistics sector seem to be worst hit by the pandemic following the lockdown order and ban on international flights and transportation as well as the mandatory closure of seaports by many economies to avert the possibility of infection by ships and shipping crew and airlines from major infected zones of the World. Following the inability of the global community so far to develop a vaccine for the cure, it has becomes imperative that the built environment, encompassing all aspects of the built environment including the environment of maritime transport and shipping must develop industry strategies, policies and regulations in line with the World Health organization (WHO) standard guidelines as preventive ways for managing and curtailing the spread of the Covid-19 disease. This is necessary to ensure that different regions of the global community does not run out of essential goods including drugs and medical equipment and food needed to sustain the fight against the pandemic and to ensure that factories involved in production of essential goods and services such as pharmaceutical companies employing imported raw materials does not shot down. The adverse effects in situation will be enormous. However, to ensure that the opening up of the seaports and the entire maritime sectors including the use of ships for operations, it must be ensured that the industry is regulated with regards to the implementation of developed Covid-19 preventive measures to guard against the infection and re-infection of settlements and society from maritime operations.

Many literatures have emerged in recent times on approaches to combating the spread of the disease in various settlement types including work places and the maritime industry. For example, the Occupational Safety and Health Academy [11], Oregon, developed a guide for protecting workplaces against Covid-19 infection. First OSHA [11] identified that the effect of Covid-19 infection in work environment such as maritime industry ranges from sickness, lockdown and exposure-fear induced absenteeism from work leading to output losses; change in patterns of commerce such as decline in consumer interest in certain goods and services as well as interrupted supply and delivery of shipments occasioned by lockdowns and cancelation of orders made from highly infected geographical regions. According to OSHA [11], workplaces must determine and different jobs with high risks of employee exposure to Covid-19 and classify each job according to the degree of exposure it holds for employing doing it. Jobs and operations should be classified as have high, medium or low risks of exposure to Covid-19 while recommendations on approaches to staying safe in each job risk category is provided to workers in line with standard guidelines. The development of an infectious disease preparedness plan is also a necessity for workplaces. Such a plan according to OSHA [11] may address adequately the need for:

i. Increased rates of worker absenteeism in high risk regions.

ii. Social distancing, adopting work shift strategy where work is staggered for individual work groups to limit the number of persons working at a given period to standard number approved by authorities.

iii. Adopting a remote work strategy where possible so that workers can work remotely from residential settlements.

These among other measures will help to prevent transmission of Covid-19 from work places to residential settlements and vice versa by reducing the risk of exposure to Covid-19.

OSHA [11] developed a hierarchy of control types based on their effectiveness for implementation in organization in the control of the spread of the Covid-19 pandemic (**Table 1**).

In another development, the Government of the People's Republic of Bangladesh [9] Department of shipping issued instructions to ports and ships on standard practices for managing and controlling the spread of the Covid-19 pandemic in the maritime and shipping sector in the country for the continued operation of ports and

Engineering control	Administrative control and safe work practices	Personal protective equipment
Isolating workers from sources of Covid-19 infection by the use of technology: i. Installing physical barriers, such as clear plastic sneeze guards. ii. Digitalization and e-commerce platforms that isolates workers from physical contacts iii. Use of robots for operation equipment with high risk of infecting human operators, etc.	These include Covid-19 work plan and policy developed by the employers in each work place in line with approved government rules to guide employee behaviors towards Covid-19 prevention outcomes. They include: i. Minimizing physical contact among workers ii. Increasing days of worker absenteeism from work iii. Developing work shifts strategy by staggering work hours for various categories of workers iv. Maximizing requirement for employee hygiene, etc.	Correct use of personal protective equipment PPE involves the use of employee protective toolkits during work to reduce the risk of indirect infection from infected objects. i. Correct fitting and compulsory use of PPE e.g. facemask, hand glove, etc. ii. Periodically refitted, as applicable. iii. PPE should be routine inspected and repaired if faulty and replaced when necessary.

Source: Modified from OSHA [11].

Table 1.
Hierarchy of control types for institutional control of the spread of Covid-19.

shipping companies. The guidelines require a ship calling to any port in Bangladesh to tender a Maritime Declaration of Health (MDH) at within 72 h before her arrival at the Port in line with the International Health regulations 2005 and FAL Conventions. The Master shall at the directive of the local Port port Health authorities provide specific information regarding health conditions such as temperature chart, crew and passenger list, current copy of ship sanitation certificate, last 10 ports call list, and list of all passengers and crew with temperatures above 37.5°C to Port Health office by email prior to the arrival of the vessel [9]. According the report, false of MDH's that does not reflect the factual conditions of health of crew and passengers onboard a ship cause the master and/or the ship agents to be prosecuted as per applicable laws. See **Tables 2** and **3** for sample of the content information in the MDH.

The report emphasized the responsibility of ship masters to educate crew and passengers of the symptoms of Covid-19, which include; (i) fever with temperature above 37.5°C/99.5°F, (ii) runny nose (iii) dry cough (iv) shortness of breath. He should also take responsibility to check the crew and passengers temperature daily, isolate confirmed cases and disinfect common areas, rooms and cabins in the ship before arrival in Ports of Bangladesh. Ports should quarantine for 14 days ships arriving from ports in infected regions following guidelines provided [9].

Passenger and crew health declaration adopt the format of listing the names and national of passengers and crew and their temperature conditions as shown below:

The GPRB [9] recommended further steps to be followed in cleaning and disinfecting ships and surfaces in the seaport to curtail the spread of the Covid-19 disease while assigning specific roles to Port Health organization, port authorities, ship owners, terminal operators, chandlers, immigration and security agencies operating ports in the duty of control and managing the spread of the disease in the Bangladesh maritime sector.

The Nigerian Center for disease Control [7, 8] provided general guidelines for mandatory institutional quarantining of returnees to Nigeria following the outbreak of the Covid-19 pandemic. Though this guideline was not specific for maritime industry, it provided framework for the 14 days quarantining of ship crew members following change of crew in ports to ensure that both the old set of crew to disembark the ship for onward movement to their families in residential settlements and the replacing crew to embark the ship are all proved to be healthy and safe, and as such cannot transmit and infect the respective new locations in residential and work environments with the Covid-19 disease. NCDC [7, 8] notes that individuals will only interact with

Name	Class or rating	Age	Sex	Nationality	Port and date joined vessel	Nature of illness	Date of onset of symptoms	Reported to a port medical officer?	Disposal of case*	Drugs, medication or other treatments	Comments

Source: Adapted from GPRB [9].

Table 2.
Sample Maritime Declaration of Health (MDH).

S/no	Name	Nationality	Temperature	Remarks

Source: Adapted from GPRB [9].

Table 3.
Sample crew/passenger declaration of health form.

approved Covid-19 surveillance officers kitted in appropriate PPE routine monitoring duties intended to facilitate the early detection of ill health due to Covid-19 and break the curve of transmission and community infection.

Furthermore, the International Chamber of Shipping [12] took necessary steps to announce guidelines for ensuring safe shipboard interface between ship and shore-based personnel including shipyards. The hierarchy of control developed by the ICS [12] is summarized below:

i. Elimination of Covid-19 hazards onboard. Example is by automating tasks and working remotely. This is the most effective approach but where it is impossible, the method of reducing the risk of infection can be employed.

ii. Reduce the risk of infection. This can be achieved by be achieved by reducing drastically the number of persons needed to carry-out the job onboard.

iii. Communicate. In a situation where the number of persons to carry-pot the job cannot be reduced and many shore-based personnel and crew must come in contact, early communication to the job-based personnel is required to get them ready on the requirements and needs of the ship and for appropriate action to be taken to ensure the protection of all parties.

iv. Personal Protective Equipment (PPE) both parties must ensure the correct and compulsory use of recommended PPE.

ICS [12] also provided steps necessary to followed by onboard crew to ensure that internal transmission and infection does not occur between and among members of crew.

Dietz et al. [13] also examined the influence of the built environment in reducing the transmission of the Covid-19 pandemic. The study harped on the risk direct and indirect spreading the virus through the built environment including the School, markets, malls hospital, recreation centers, religious worship centers etc. outlining necessary action plan for routine disinfection of objects and control of person-to-person transmission of the disease in the built environment.

Finally, the International Maritime Organization harped on the need for a common framework and protocols for joining ship from ordinary residence in one country via aircraft to ship ship in a seaport in another country as well as a common protocol for leaving a ship and repatriation from a seaport in one country via aircraft to a seafarer's place of ordinary residence in another country. This was to address the challenges faced by ship-owners in changing crew following the outbreak of the Covid-19 pandemic. Further provisions were issued to coastal states, shipping companies and agents, allied organization by the IMO [14] noting the importance of granting seafarers movement pass and access to travel facilities as provider essential services by national governments.

From foregoing, literature gaps were identified which among other things include the fact many organizations such as the WHO, NCDC, Departments of shipping, ICS, IMO, national governments, etc. have issued rules, recommendations, protocols and guideline for curtailing the spread of the Covid-19 disease in the maritime industry without identifying clearly a responsible organization in the maritime industry with responsible to monitor implementation of the guidelines. A holistic approach has not being followed in developing most of the preventive frameworks suggestive for the maritime industry, for example, the IMO [14] preventive framework centered more on change of crew and the role of government and other organization without considering among other things the transmission and infection of ships by seaports and from interactions with residential settlements. The WHO [5] and NCDC [7, 8] prevention and 14 days quarantining measures were is for generality of all institutions and lacked the consideration of the peculiarity of the maritime industry, thus maritime industry specific measures need to be developed but in conformity with WHO and NCDC guidelines and recommendations. The routes and possible channels of transmission and infection of ships and residential settlements as it affects the maritime industry based in the interactive relationship among stakeholders in the maritime industry and residential settlements which holds potentials to transmit the virus and well possibility of infection need to be developed as basis for assessing the risk of exposure and transmission of the disease so that the various WHO, NCDC, IMO and other industry frameworks, protocols, rules and guidelines for preventive the spread of the Covid-19 disease can be effectively implemented along the routes/channels of transmission. It is the above identified literature and knowledge gaps that the chapter has tried bridge in the earlier sections.

4. Challenges faced in harnessing the built environment for prevention of the spread of Covid-19

It is pertinent to mentioned that it is important to harness and built the capacity of the entire built environment of maritime transport and other modes of transport to contribute to the fight against the transmission of the Covid-19 pandemic until an approved vaccine is developed for the disease. However, there exist a plethora of challenges to fast tracking of the processing and inputs towards improving the capacity of the maritime industry to actualized the dream of limiting the spread of the disease. Some of these challenges include:

i. Financing the implementation of the preventive measures poses a challenge. Financial limitations and challenges associated with accessing funds needed to overcome the cost implications of implementing the maritime industry general and company specific guidelines for the prevention of the spread of the disease in line with the WHO and NCDC guidelines. This is because the Covid-19 pandemic currently has negatively impacted the economies of most maritime companies and the allied organizations following series of lockdowns over the past months, leading to output losses. Since implementing these preventive guidelines (both general, industry specific and company specific guidelines) requires funds. For example, the cost of acquisition of sanitary wares and engineering designs as guides against the spread of the disease has increased over time. This is in the face of lockdown associated inactivity and output losses in many sectors of the economy. Therefore, financing the implementation of the preventive regulations and guidelines in the environment of maritime transport poses a major challenge particularly for smaller companies and new start ups.

ii. Shift and/or change in work practices may affect productivity levels at least in the short run. So Covid-19 preventive guidelines adopted in managing its spread in the work environment such as the strategy of staggering work hours for different employees in a given firm and redesigning the work environment to conform Covid-19 preventive plans may lead to shift (decline) productivity at least, in the short run. Thus this will influence the decision of what guidelines to fully implement and to what extent by firms.

iii. Implementing the Covid-19 preventive guidelines in the wide maritime industry is important. It will however imply a drastic behavioral change aimed at eliminating risky behavior. This will in turn influence and/or require in change in worker attitude towards and during work which cannot be instant. Thus training and retraining on the best alternatives action plans to prevent the spread of the disease in necessary and the absence of the capacity to train and retrain with the associated patience needed to get employees develop required work non-risky worker behavior is a challenge [15].

iv. Finally, time constraint also poses a challenge since the rate of spread of the pandemic requires immediate action by relevant stakeholders in combating it. Time lag in implementation of for example, engineering designs as recommended by OSHA [11] may increase the spread of the disease within the industry and the interacting settlements and built environment types.

5. Recommendations

1. From the foregoing, it is obvious that the maritime industry does not constitute solely of ships, port authorities and shipping companies. Therefore frameworks for prevention of Covid-19 transmission solely from the perspectives of the trio cannot sustainably guarantee the safety of the industry and the allied sectors from the disease. All frameworks for the prevention of the transmission in the maritime industry must be holistic enough to encompass all the allied sectors, including but not limited to customs and other government agencies, freight forwarders and multimodal transport operators, ship chandlers, shipyards, and all stakeholders and sub-components of the maritime sub-sector.

2. All components and sub-components of the maritime industry and operators must commence by first identifying based on their specific model of interaction the likely internal and external routes/channels of spreading the Covid-19 disease within, to and by the organization. Specific company and/or industry regulations and guidelines cum the general standard guidelines can now be implemented following the identified likely channels of transmission in line with WHO and local center for disease control (CDC) guidelines.

3. Exposure risk assessment should form the first basic task to be carried by all industry stakeholders to determine the level of exposure while standard guidelines can now be subsequently applied.

Author details

Theophilus Chinonyerem Nwokedi
Department of Maritime Management Technology, Federal University of Technology, Owerri, Nigeria

*Address all correspondence to: nwokeditc@gmail.com

References

[1] Assari A, Birashk B, Nik M, Mousavi Naghdbishi R. Impact of built environment on mental health: Review of Tehran City in Iran. International Journal on Technical and Physical Problems of Engineering. 2016;**8**(26):81-87

[2] Roof K, Oleru N. Public health: Seattle and King County's push for the built environment. Journal of Environmental Health. 2008;**75**:24-27

[3] Nwokedi TC. Covid-19 pandemic and the implications on container shipping trade in Nigeria. In: Seminar Presentation at Atlantic Maritime Academy Container Trade Seminar Series, RUST Campus, Onne Seaport, Nigeria. September, 2020. 2020

[4] Heath G, Brownson R, Kruger J. The effectiveness of urban design and land use and transport policies and practices to increase physical activity: A systematic review. Journal of Physical Activity and Health. 2006;**3**(s1):S55-S76. DOI: 10.1123/jpah.3.s1.s55

[5] World Health Organization (WHO). Rolling updates on corona virus disease (COVID-19). 2020. Available from: http://www.who/int/countriesi/nigeria [Retrieved: May 23, 2020]

[6] Nallon E. COVID-19: A Maritime Perspective. 2020. Available from: http://www.maritime-executive.com [Retrieved: May 22, 2020]

[7] Nigeria Center for Disease Control (NCDC, 2020) National Interim Guidelines for Clinical Management of COVID-19, version 2, May 2020. Available from: http://www.Covid19.ncdc.gov.ng [Retrieved: May 23, 2020]

[8] NCDC. Mandatory Institutional Quarantine Guideline for Returnees to Nigeria. May, 2020. 2020. Available from: http://www.ncdc.gov.ng/covid19 [Retrieved: May 26, 2020]

[9] Government of the People's Republic of Bangladesh (GPRB, 2020) Instructions to ports and ships for dealing with corona virus (covid-19) pandemic. Circular Number: 02/2020. Issue Date: 25-03-2020. Available from: http://www.dos.gov.bd [Retrieved: May, 26, 2020]

[10] Okon JS. Safety and Security Aspects of Maritime Transportation. Uyo: Modern Business Press Ltd; 2012

[11] Occupational Safety and Health Academy (OSHA). Guidance on Preparing Workplaces for COVID-19. Oregon, USA: OSHA Academy; 2020

[12] International Chamber of Shipping (ICS, 2020) Covid-19 Related Guidelines for Ensuring a Safe Shipboard Interface Between Ship and Shore-Based Personnel. Available from: http://www.iCs.org/Covid-19 [Retrieved: May 23, 2020]

[13] Leslie D, Patrick FH, David AC, Mark F, Jonathan AE, Kevin VDW. 2019 Novel Corona virus (COVID-19) Pandemic: Built Environment Considerations to Reduce Transmission. Applied and Environmental Science. 2020. Available from: https://www.researchgate.org

[14] International Maritime Organization (IMO, 2020) Corona virus (COVID-19)—Recommended framework of protocols for ensuring safe ship crew changes and travel during the corona virus (COVID-19) pandemic. London. Circular Letter No. 4204/Add.145. May. 2020

[15] Andrei RA, Kenji M, Sung-mok J, Natalie ML, Ryosuke O, Hiroshi N.

Estimation of the actual incidence of coronavirus disease (COVID-19) in emergent hotspots: The example of Hokkaido, Japan during February–March 2020. 2020. Available from: http://www.researchgate.org [Retrieved: May 02, 2020]

Chapter 4

Principles of Sustainable Logistics

Baha M. Mohsen

Abstract

Effective logistics largely contributes to the success of business through quick deliveries in minimum time and cost. Logistics is the process of getting material, product, and service where and when they are needed. When addressing the concept of sustainable logistics, it is important to consider the three dimensions of sustainability: economic, social, and environmental. Sustainable logistics is beyond going green and being environmentally friendly, it has an influence on manufacturing processes, starting from where raw materials are obtained, processes involved, use, and potential recycling of the product or service. In analyzing the problem of evaluating logistic operation performance, sustainability will be one among other criteria for evaluation. Selecting a logistics partner who cares about sustainability will help in achieving company's strategy. The chapter discusses the concept of sustainability applied to logistics. From the main definition of the logistics system, which includes the flow of materials and information, to the goal of sustainable logistics, which includes increasing profitability and reducing the environmental impact. Sustainable development involves coordination between environment and economics to reach social, economic, and environmental sustainability. The chapter aims to help managers, practitioners, scholars, and students to understand the key attributes of sustainable logistics and supply chain in general.

Keywords: logistics, sustainable logistics, green logistics, sustainable development, multi-criteria decision making, environmental sustainability, economic sustainability, social sustainability

1. Introduction

Every aspect of our lives includes some form of supply chain and logistics, so the impact of these activities on the environment is of significant importance. The objective of this chapter is to introduce principles and practices that facilitate sustainable logistics operations in a holistic manner and consider factors of logistics affecting the natural environment beyond the usual factors of distance traveled, fuel use, and carbon dioxide (CO_2) emissions that have been well discussed in freight transportation literature. Global business is more dominant these days, for example, many manufacturers produce their goods in lesser-developed countries, and then shipped all around the world. This requires global logistics to ensure timely and efficient global distribution of goods from producers to consumers. Logistics and supply chain management

(SCM) activities have a significant economic impact on countries and their societies. Grant et al. [1] reported these activities accounted for 8.3 percent of US gross domestic product (GDP) or US $1.45 trillion in 2014 and 6.8 percent of GDP (€876 billion) across the European Union's (EU) 27 countries in 2012.

This chapter covers sustainable logistics. It starts by presenting an overview of logistics in the second section. The third section discusses the basics of sustainable development and sustainability. The fourth section introduces the basic concept of sustainable business as minimizing costs, which covers the three aspects of business: environmental, financial, and human. The fifth section discusses sustainable supply chain and logistics. As business is becoming more challenging these days, companies need to be aware of and practice sustainable supply chain management to stay competitive. Sustainable supply chain management is about environment protection, social responsibilities, economic growth, and profitability in the long term. It wraps up with a brief section on the evaluation of logistic operations. In addition to cost and speed criteria, sustainability should be introduced in the evaluation criteria.

2. Logistics

Effective logistics largely contributes to the success of business through quick deliveries in minimum time and cost. Logistics is the process of getting material, product, and service where and when they are needed. It works to determine the temporal and spatial positioning of raw materials, work in progress, and finished inventories where they are needed and when they are required. Logistics can be categorized into subsistence logistics, operation logistics, and system logistics. Subsistence logistics is concerned with the basic human needs of food, clothing, and shelter within any given conditions, and it provides the foundation of operations logistics. Operations logistics goes beyond subsistence to systems involved in producing luxuries; it incorporates the raw material required by the enterprise in the production. System logistics includes all resources required in keeping a system in operating condition. These resources include personnel, test and support equipment, spare parts for maintenance, technical publication, and facilities. Thus, logistics systems consist of four main activities: purchasing management, inventory management, warehousing management, and transportation management.

Logistics is defined by the Council of Logistics Management (CLM) as "the process of planning, implementing and controlling the efficient, effective flow and storage of goods, services and related information from the point of origin to the point of consumption for the purpose of conforming to customer requirements" [2].

Effective logistics minimizes the cost of transportation, inventory, material handling, and other distribution-related activities. In light of the new trends in business, logistics have gained great importance. New trends include high production efficiency, change in inventory philosophy, high transportation cost, production lines replication, propagation of computers and technology, retails fast-growing, globalization, and reduction in economic regulations. Efficient logistics systems throughout the world business are a basis for trade and a better economy. It allows a geographical region to exploit its inherent advantage by focusing its productive efforts on those products in which it has been an advantage, which will result in competitive production cost, logistics cost, and quality compared to other regions.

Global logistics is growing and playing a vital role in international business. It ensures timely and efficient global distribution of goods from producers to consumers through a connection of critical components of the supply chain from a product's point of origin to its point of consumption. It was reported that global container trade has increased on average 5 percent per year over the last 20 years and at its peak in the mid-2000s comprised 350 million 20-foot equivalent units (TEU) a year [1].

Advancements in information technology and communication, transportation and material handling, and high volume data processing and transmission are revolutionizing logistics control systems. The use of big data tools in logistics and supply chain management gives great advantages as it provides better decision making, improved efficiency, cost reduction, better risk management, and better visibility and competition [3]. Communication technology enables better, faster, and reliable supply chains, communications take place between any firm, suppliers, customers, and other members involved in the chain.

3. What is sustainability?

Different people or organizations might have different understandings or definitions for sustainability, depending on their area of specialty or function. Many people think the word sustainability is synonymous with "going green," or limit their understanding of sustainability to the environment. However, the word sustain means causing or allowing something to continue over a period of time. In the same logic, an unsustainable process or act assumes that it will come to an end sooner rather than later. The World Commission on Environment and Development (WCED), known by the name of its Chair, Gro Harlem Brundtland, published its report "Our Common Future," in 1987 [1] and proposed the concept of sustainable development as an ideal for the global economy and corporations. Sustainability was defined as development that meets the needs of the present generation without compromising the ability of future generations to meet their needs. Based on the WCED definition of sustainability, focusing only on the environmental aspects of sustainability is short-sighted and partial. In addition to the environment, sustainability embraces several arenas including economics, materials, industry, human behavioral science, laws and legislation, social sciences, and finance, as depicted in **Figure 1**.

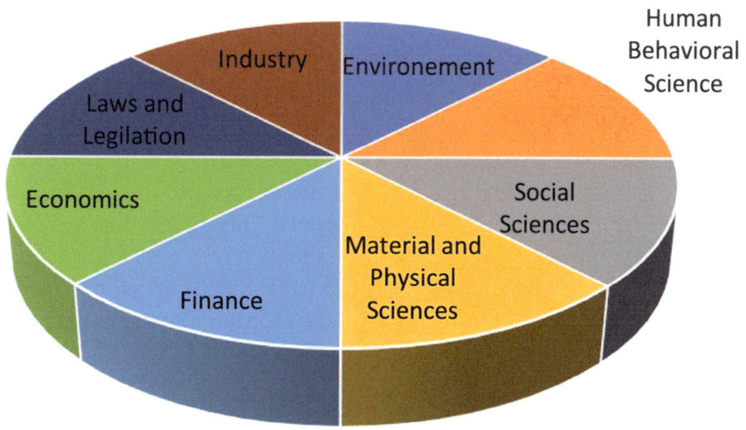

Figure 1.
Arenas of sustainability.

To achieve sustainable development, there are three goals that need to be fulfilled: waste elimination or minimization, optimization of resources, and cost minimization. Achieving the three goals will lead to environmental sustainability, economic sustainability, and social sustainability, which are the three pillars of sustainability. From a business viewpoint, sustainability is about reducing costs in every conceivable form, which will lead to profitability, competitiveness, and continuity. These costs consist of the costs of raw materials, waste, deficits in resources, poor product design, inefficient production process, climate change, and unemployment. These costs can be grouped into three categories that represent the three aspects of business: Environmental, Economic, and Social.

4. Sustainable business

Sustainable business is about minimizing costs, which covers the three aspects of business: environmental, financial, and human. There are several challenges business need to consider to better compete in the market. The volatile energy prices is a major challenge for business, all options need to be considered to reduce the energy bill and consequently the business's dependence on oil. The prices of raw material increases due to increase in the world's population. Increases in waste and disposal costs are becoming critical to business. Most countries of the world have passed laws and regulations for waste regulation and recycling. Plastic, cans, papers, and other recyclable materials are kept away from landfill sites to avoid waste. The legislation to control CO_2 pollution is active in many countries, manufacturers and industrial companies not only face a real challenge to keep their level of CO_2 within limits, but investors are avoiding investing in companies that are not willing to comply and watch the high cost of change in environment. Customer and business demands and expectations are becoming different, buyers are so much aware of prices and sustainability initiatives. Retailers are concerned with their supply chains, where the majority of their environmental footprint is centered. They are interested to optimize prices of energy, material, and supply chains. Companies that are committed to low-cost sustainable operations gain the best market share; this will put such companies on a competitive advantage compared to others. Nowadays, customers expect companies to be transparent, which is done through two channels: voluntary where information passing from company to customers, or involuntary where information is shared by consumer watchdog groups. There are companies that are making it possible for customers to have online access to follow up on products from concept to material sourcing, to manufacturing and delivery. Companies share ingredients of their products by providing online lists.

Commitment to sustainability helps companies to recruit, and retain smart employees who think about things other than money. Those employees work with pride and purpose, want to feel the ability to make difference and accomplishment. The longer a company takes an action, the higher the cost is, and the further behind it will be in terms of market share, profitability, and innovation. Hidden costs exist, such costs could be building-related low productivity, and sickness resulting from poor ventilation and lighting. In addition, there are costs related to laying off employees, which include the loss of investment in human capital, economic and social ex-communication, and reduction in national economic growth.

5. Sustainable supply chain and logistic

As business is becoming more challenging these days, companies need to be aware of and practice sustainable supply chain management to stay competitive. Sustainable supply chain management is about environment protection, social responsibilities, and economic growth and profitability in the long term. **Figure 2** depicts the relationships between the three goals that need to be fulfilled: waste elimination or minimization, optimization of resources, and cost minimization, sustainable development with its three dimensions: economic sustainability, environmental sustainability, and social sustainability, and sustainable logistics with its components: logistics concepts, methods, and functions. To achieve sustainable development, integration of its three dimensions is required; any defect in the three dimensions of sustainable development will not lead to its achievement. Logistics is involved in all aspects of business as well daily life of individuals. Sustainable logistics is tied to sustainable development in general, sustainability criteria should be included in the logistic evaluation, in addition to another criterion such as cost and speed. Sustainable logistics is at the intersection of its concepts, methods, and functions. The goal is to eliminate environmental problems in the areas of logistics, which can be achieved by eliminating or minimizing negative impacts of logistics on the environment. Starting from the concepts, these activities include designing sustainable packaging, and reuse of, recycling waste, reducing energy and the pollution caused by transport.

Several concepts and terms of logistics resulted from strict environmental regulations. Reverse logistics is defined as "the process of planning, implementing and controlling the efficient and cost-effective flow of raw materials, semi-finished and finished products, together with the related information flows, from the point of consumption to the point of origin, in order to recover the value or proper

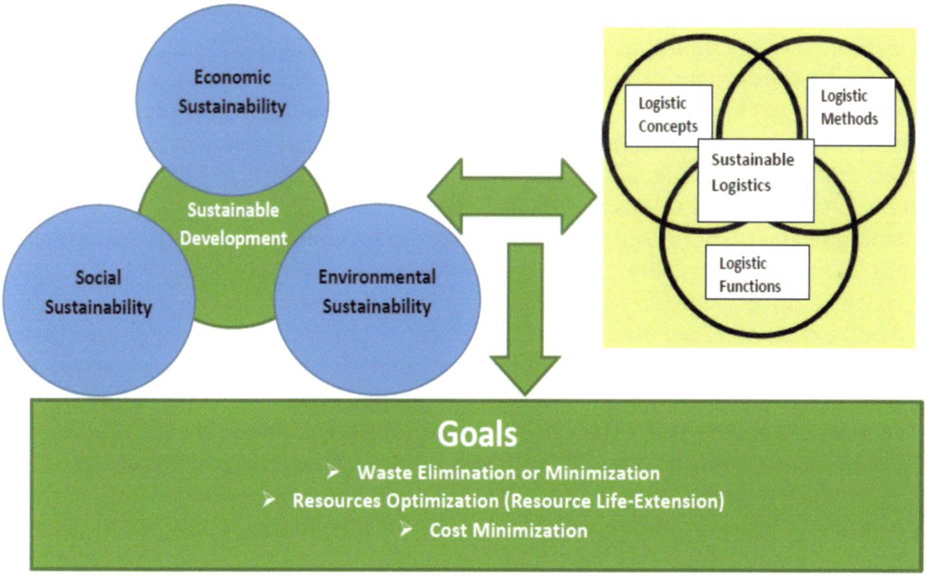

Figure 2.
Sustainability framework.

management" [4]. Disposal logistics – "the application of the concept of logistics for the residue, to induce their efficient, economically and ecologically, movements, while the space-time transformation, including changes in the amount and type of" [5]. Recirculation logistics – "suggests that product or packaging, is circulated repeatedly in a closed-loop supply chain" [6]. Downcycling – "process of waste or useless products transformation into new materials or products, having lower quality and functionality" [6]. Green logistics is defined as a "management approach" aimed at minimizing the negative impact on the ecosystem of logistics flows. The problems of excessive environmental degradation concern companies, operating in each market sector. In particular, however, apply to freight forwarders and carriers. The concept of green logistics associated with the strategy consists in the use of their resources in the most efficient and environmentally friendly way. It is a trend that stems from the need to care for the global environment [7].

Logistics includes "efforts to acquire materials and finished products distribution to the right place, at the right time and in the right amount. Typical elements of the logistics system are customer service, demand forecasting, distribution communications, inventory control, warehousing, procurement processes, parts and service support, site selection magazine, shopping, packing, handling complaints, waste management, transport, and storage" [8]. To improve logistic operations, companies collaborate with suppliers, shippers, distributors, and customers. As a result, logistic cost will be reduced and business performance will be improved.

Sustainable supply chain management covers all activities, functions, processes, and relationships, where materials, products, services, information, and monetary transactions move among enterprises. The first step in the implementation of sustainable supply chain management starts with product design. In addition to optimizing quality and cost, the design will allow recycling of products. Sustainable production is the second step, which can be achieved through utilizing clean production methods, use of new technology, reducing raw materials, and resources. Sustainable marketing helps companies to enhance their relationships with stakeholders. Maintaining biological balance, paying more attention to environment, and waste management leads to cost reduction and improved competitiveness. Sustainable transportation is a major element in achieving sustainable supply chain management. Utilizing renewable energy, modes of transport, infrastructure, and operational management practices can be considered to achieve sustainable transportation. Sustainable purchase leads to minimizing waste, hazardous materials, and sources of pollution.

Hammer and Somers [9] discussed concepts that provide possibilities of using resources more productively. The lean methods involve following a product through factory or service operation with the objective to reduce waste of energy and materials. Unlike profit per ton, the concept of profits per hour takes into account the time dimension in production process. This concept enables companies to make wise decisions and choices regarding resources and productivity.

Advanced analytic techniques help companies navigate and sort within different variables such as equipment configuration, raw materials, and process changes. Comprehensive change management effort is required for resource productivity, which ensures that employees create more value from less. Think circular is a sustainable logic that creates new value for companies and societies. This logic relies on looping products, components, and materials back into the production process.

Rothenberg [10] promoted the concept of "servicizing," where suppliers could focus on providing services instead of selling products as their business models.

This will lead to reduced material use as a strategic opportunity. This is in line with the World Commission on Environment and Development definition of "sustainable development." The author presented the case study of three companies; Gage Products, PPG Industries, and Xerox. The three companies are taking the servicizing approach; they adopted business models that help customers purchase less of their products. The three companies have attracted new customers with their new business models. In addition, they have built stronger customer relationships. This closure customer relationship has led to expanding the range of products they sell, attracting new customers who are impressed with the company's sustainable social commitment, and usually, customers are less likely to change suppliers.

Similar to other large-scale initiatives for change, moving to servicing is faced with challenges. Changing the culture from selling more products to helping customers to use fewer products is not an easy task. There will be internal resistance at different levels to this change. Managing this change falls into six categories: (1) utilizing existing strengths (2) restructuring profit in contractual agreements (3) introducing the new business model (4) new incentive schemes (5) introducing new skills and (6) expressing major significance and special interest of environmental advantage.

Paine [11] discussed the importance of corporate responsibility to their long-term success. Nike's corporate committee's role in supporting innovation was described. The committee was created in 2001 as a result of board member Jill Ker Conway's lobbying. The committee advises on issues such as labor practices, resource sustainability, innovation, and acquisitions. The author concluded that corporate responsibility committee could help companies in several ways such as a source of knowledge and expertise, accountability, driver of innovation, a source for the full board and constructive critic.

Nidumolu et al. [12] discussed the idea of collaboration, as is a necessity for business sustainability. They introduce four models for systematic sustainability using case studies. The models have two common features; stakeholder inclusion and innovation in either operating processes or business outcomes. Companies can work together on issues such as climate change, depletion of resources, and ecosystem. Two types of collaborations focus on business processes and outcomes. First, is corporate collaboration, which includes all players in business such as manufacturers, suppliers, distributors, and retailers. Second, extended collaborations, which covers business and non-corporate partners such as government, NGOs, and academics.

Companies can identify and share operational processes that will minimize consumption of resources and waste, which will lead to natural resources protection. In addition, companies can share defined outcomes that minimize environmental impact. To explain the proposed models, the authors discussed several case studies. Processes were the center of corporate collaborations in the case of Dairy Management Inc. and case of an action to accelerate recycling. The corporate collaboration's focus was on outcomes in the case of sustainable coalition, and Latin American water funds partnership.

Authors stated seven next practices for successful collaborations in sustainability. (1) Starting with small committed group (2) link self-interest to shared interest (3) monetize system value (4) create a clear path with quick wins (5) acquire independent project management expertise (6) build in structured competition, and (7) nurture a culture of trust.

Doorey [13] describes the case study of Nike and Levis. The contribution of factory disclosure was evaluated, and tracking the change from resistance to transparency in supply chain was tracked. Information disclosure is a tool that is used in business and it affects its behavior. Business leaders can change harmful behavior with transparency and empowering private watchdogs. This is clear from the private movement that took place in the late 1990s to pressure corporations to disclose and declare their global suppliers. It was believed that such disclosure would improve working conditions and labor practices. For example, Nike and Levis published their suppliers list in 1995, which was a surprise to the business community. Information systems were introduced to track information about labor practices including global suppliers' databases.

Unlike Nike and Levis, many other companies are not welcoming the idea of supplier transparency. Some of these companies cannot track their suppliers, which will lead to ignorance of labor practices. In such case, the role of private actors becomes essential to apply the needed pressure to create transparent supply chains.

New [14] discussed the importance of transparent supply chain. When supply chain is not transparent, trouble will arise. Many companies consider provenance very important, H&M for example declares that labor practices and environmental effects of its suppliers' suppliers are very important. Origin is considered an important and essential feature of what a customer may buy. Companies such as Wal-Mart are using new technologies to provide origin data. For example, bar codes that can be read with mobile phones, genetic markers for agricultural products and labeling have been transformed by microscopic electronic devices.

Customers have an interest in origins and authenticity. Providing them with information about provenance will become part of the marketing strategy. Provenance is important on downstream and upstream sides of the supply chain. Both customers and suppliers can access this internet of things that gather provenance information.

6. Evaluation of logistics

Until recently, logistic operations were evaluated based on cost and speed criteria. The situation is different now, where sustainability was introduced in the evaluation criteria [15]. Logistics is a human activity and impacts the environment through its components purchasing, inventory, warehousing, and transportation. Decision-making becomes more important in logistic and supply chain management, which may vary from single quantitative criterion to multi-criteria problems. Many problems in logistic management such as supplier selection, purchasing, manufacturing, distribution, collaboration, performance management, and design are covered by suitable multiple criteria decision-making approaches. Mohsen and Murat [16] developed an analytical hierarchy process (AHP) methodology to perform a multi-criteria evaluation of freight forwarders. It analyzes different criteria that would be employed in the evaluation and selection of forwarders.

Companies' competitiveness rely on their ability to ship their products around the globe at the right time in a perfect physical situation. It is suggested that logistic evaluation is based on a multi-criteria in terms of transit time, quality of service, expertise and specialized service, network, competitive prices, technology and information, and sustainability criteria.

6.1 Quality of service

In logistics, quality of service is not about responding to emails or calls anymore. The quality of service pertains to such areas as expertise in providing relevant services, ability to work with one point of contact, meeting unique customer requirements, dependability and assurance of the international shipping service, flexibility and ability to provide a wide range of services, and meeting deadlines. Quality of service is about supporting customers to gain a competitive edge against their competitors by using innovative systems, utilizing big data, and advising on the market conditions and logistics trends. Huang et al. (2019) explore some practical business solutions to enhance customer service level of the international freight forwarders. Providing high-quality service is a key objective in this business sector to enhance customer satisfaction since competition is extremely critical.

6.2 Specialized service

Ability to provide specialized service is very critical for customers needing special and expert handling of their sensitive cargo. For example, the shipment of a pharma product requires a forwarder who knows the regulations and has proven record experience in handling pharma when it comes to temperature monitoring during the shipping and warehousing requirements.

6.3 Network

Network for the logistics partner is defined as the existence of the company around the globe. It is important to deal with a company that has a strong network around factories to the distribution centers to make sure they guarantee space for cargo, and on the other hand to be committed to the agreed transit time. Network dimension concerns such factors as international deployment, number of branches worldwide, and number of countries in which the international shippers are represented.

6.4 Competitive price

Price might be critical to some industries, especially for low-value products. Hence, customers seek the most competitive rates in the market to increase their profitability. On the other hand, other industries' shipping pricing might not be as critical as for pharma and high technology because of the need for special handling and expertise which might not be available in all logistic companies. So, you need to check your industry and value of your product before negotiations. In all cases, the ability to offer service with attractive prices will continue to be an important criterion.

6.5 Information and technology

No doubt that technology is so important nowadays, finding a partner with advanced technology would help to have efficient and effective operations. The ability to access and interface with the international shippers' information technology is a very important factor in evaluating the freight forwarder. Finding a partner with advanced technology will give the company a competitive edge against your competitors by utilizing their big data.

Some shippers have really advanced technology to level that they have platforms, which allow you to predict the economy for every quarter of the year per region by utilizing the big data of logistics (import & export), this will be very helpful, especially for newly launched products.

6.6 Sustainability

Saving the environment is everyone's concern nowadays, all companies giving more attention to sustainability and they made a lot of initiatives to save environment. Finding a logistics partner who cares about sustainability will help the company in its strategy. Some of the logistic and transport companies generate reports for CO_2 emission, which will help you in your strategy.

7. Conclusions

Logistics is a required function for all types of businesses. It covers many actions and activities performed by the companies involved in managing the flow of raw material, unfinished products, and final products. This wide range allows to introduce and use many tools, solutions, or actions that led to the creation of Sustainable Logistics Management term.

Sustainability is the future for logistics and supply chain businesses. It is important to understand the level of social, environmental, and economic impact and viability that suppliers and customers have. It is beyond going green and being environmentally friendly, it has an influence on manufacturing processes, starting from where raw materials are obtained, processes involved, use, and potential recycling of the product or service. In analyzing the problem of selecting a freight forwarder, sustainability will be one amongst other criteria for evaluation. Selecting a logistics partner who cares about sustainability will help in achieving company's strategy. Some of the logistic and transport companies generate reports for CO_2 emission, which will help in achieving specific targets and strategies.

As business is becoming more challenging these days, companies need to be aware of and practice sustainable supply chain management to stay competitive. Sustainable logistic and supply chain management is about environment protection, social responsibilities, and economic growth and profitability in the long term.

In addition to cost and speed criteria, sustainability should be introduced in the evaluation criteria. It is suggested that logistic evaluation is based on a multi-criteria in terms of transit time, quality of service, expertise and specialized service, network, competitive prices, technology and information, and sustainability criteria.

Principles of Sustainable Logistics
DOI: http://dx.doi.org/10.5772/intechopen.103018

Author details

Baha M. Mohsen
Industrial and Systems Engineering Department, Wayne State University, Detroit, Michigan, USA

*Address all correspondence to: baha.mohsen@wayne.edu

References

[1] Grant DB, Trautrims A, Wong CY. Sustainable Logistics and Supply Chain Management: Principles and Practices for Sustainable Operations and Management. 2nd ed. Vol. 10. New York: Kogan Page Ltd; 2017

[2] Council of Supply Chain Management Professionals. n.d. CSCMP Supply Chain Management Definitions and Glossary, Council of Supply Chain Management Professionals. [online] Available from: https://cscmp.org/supply-chain-management-definitions [Accessed: 7 December 2022]

[3] Mohsen B. Big data applications in supply chain management: SCOPUS based review. In: 6th International Conference on Information Systems Engineering (ICISE 2021). Shanghai, China, New York, NY, United States: Association for Computing Machinery; 2021

[4] Rogers Dale S, Tibben-Lembke Ronald S. Going Backwards: Reverse Logistics Trends and Practices. Nevada: Reverse Logistics Executive Council; 1998. p. 125

[5] Litman T, Burwell D. Issues in sustainable transportation. International Journal of Global Environmental Issues. 2006;**6**(4):331-347

[6] Meiling He ID, Shen J, Wu X, Luo J. Logistics Space: A Literature Review from the Sustainability Perspective. Sustainability. 2018;**10**:2-24

[7] Al-Odeh M, Smallwood J. Sustainable supply chain management: Literature review, trends, and framework. International Journal of Computational Engineering & Management. 2012;**5**(1):86

[8] Bajdor P, Grabara I. Green supply chain in the enterprise operating at the fashion market. Logistyka. 2012;**6**:3-18

[9] Hammer M, Somers K. More from less – Making resources more productive. McKinsey Quarterly. 2015:1-7. Available from: https://www.mckinsey.com/business-functions/operations/our-insights/more-from-less-making-resources-more-productive

[10] Rothenberg S. Sustainability through servicing. MIT Sloan Management Review. 2007;**48**(2):83

[11] Paine LS. Sustainability in the Boardroom. Harvard Business Review. 2014;**92**(7):86-94

[12] Nidumolu R, Ellison J, Whalen J, Billman E. The collaboration imperative. Harvard Business Review. 2014;**92**(4):76-84

[13] Doorey DJ. The transparent supply chain: From resistance to implementation at Nike and Levi-Strauss. Journal of Business Ethics. 2011;**103**(4):587-603

[14] New S. The transparent supply chain. Harvard Business Review. 2010;**88**:1-5

[15] Muller J. Delivering Tomorrow Toward Sustainable Logistics. Deutsche Post Ag: Bonn, Germany; 2010. p. 82

[16] Mohsen B, Murat E. Multi-criteria selection of a freight forwarder using analytic hierarchy process. International Journal of Analytic Hierarchy Process. 2021. Submitted for publication

Section 3

Inventory Control

Chapter 5

Advances in Inventory Control

M. Rameswari

Abstract

Inventory control is an important part in logistics, which could manage the inventory by making revenue to keep continuous moving on to strengthen logistics and inventory management activities. The desired results and ensure measurable targets are set to drive the right decisions across the organization. The objective of the proposed model is to take into account the distributed block chain ledger system in industry. Block chain technology on inventory management will become a good traceability system. The impact of such digitization will be a very effective and efficient process that facilitates transactions for all industry people in the network.

Keywords: inventory, cost of inventories, production inventory, backorder inventory, inventory management, block chain technology

1. Introduction

Inventory control plays an important role in running a business smoothly and ensuring a better economy for the company. Inventory is in the form of raw material, finished goods, or semifinished goods, etc., which are supplied to the customers at right time according to their need. A business needs to implement inventory control so that it can produce an optimum level of inventory and avoid excess inventory, minimize loss due to obsolescence, deterioration, and meet demand fluctuations. Inventory control is the process of ensuring to keep surplus, scrap, and obsolete at the minimum level, minimizing the shortage is due to nonavailability of stock at the right time, and with minimum holding cost.

2. Inventory control

Maintaining inventories is one of the most important issues in logistics supply management because it keeps the physical presence of the materials for future use. Inventory control in the industry is the process of maintenance of the stock to meet customer demand in terms of quantity and quality. The industry must have good inventory control, which is a common challenge in making the right decision for effective inventory management. Inventory management supervises the flow of goods from manufacturers to warehouses.

Inventories are stock piles of raw material, finished goods, or semifinished goods, etc., used in organization. On the other hand, in some industries such as high-tech industries, it is noted that components' cost is decreasing at a sustained rate.

Due to the fluctuations in demands and lead time, the industry has to make decisions depending on the inventory.

1. How much to order?

2. When to replenish?

2.1 Cost of inventories

There are different types of costs to maintain in inventory, which are discussed in the following.

2.2 Ordering cost

The fixed cost of order quantity includes the postage, advertisements, consumption of stationary items, telephone charges, transportation, and incurred traveling expenditure and so on.

2.3 Holding cost

The cost associated with holding the goods in stock is called the holding cost. This cost includes the costs such as rent for space used for storage, facility maintenance, and related costs, insurance, taxes, breakage, pilferage, communication costs, and utilities, besides the cost of human resources employed in operations as well as management and warehouse overhead also depend on the value of the inventory.

2.4 Shortage cost

The cost associated with the items for running out of stock. That is, there is no inventory in stock. This cost includes the loss of potential profit through sales of items and loss of goodwill. It will cause the permanent loss of customers, which may cause the lost profit in future sales.

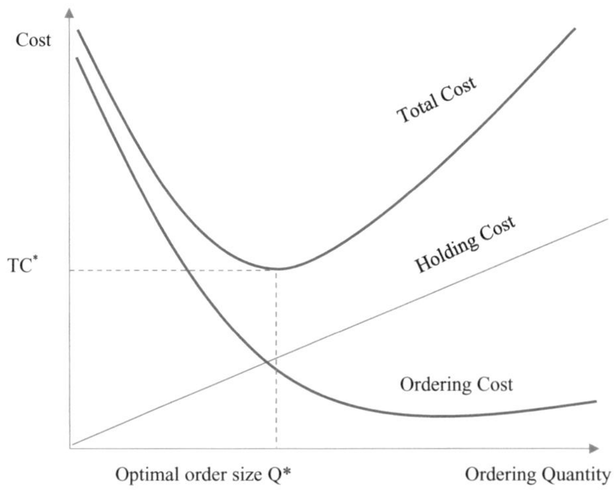

Figure 1.
Graphical representation of optimal order size and minimum total annual cost.

2.5 Revenue cost

The expensive cost in inventory incurred when an organization earns revenue by selling its products/services to the customers. It is the total cost of production and delivering a product or service to consumers. See Ref. [1]

If number of orders is increased, the ordering cost will be high, but the holding cost may be low. If the number of orders is decreased, the ordering cost will be low, but the holding cost may be high. Optimal order quantity is that order quantity that minimizes the total annual cost of ordering inventory and cost of holding inventory. See Ref. [2]. If the ordering cost and holding cost are equal, then the total annual cost is minimum as shown in **Figure 1**.

3. Important factors in inventory control

3.1 Demand

The number of units required per period is called demand. The demand pattern of a commodity may be either deterministic or probabilistic.

3.2 Lead time

The time gap between placing of an order and its actual arrival in the inventory is known as lead time. The inventory level of an item depends upon the length of its lead time.

3.3 Deterioration

A change or loss in the original value of the product is caused by certain factors that can worsen or damage the product. This is known as deterioration. It is common in goods such as vegetables, food products, dairy products, fruits, pharmaceutical products, etc. These items have less life time due to deterioration process.

Some products such as dried fish, usage of which may increase over time in an inventory level, or some products such as fast-growing animals by which its value in quantity to make better or increase their inventory level is known as amelioration. In stored items, the inventory may be affected by degrading or may be affected by upgrading their value to increase their size or increase their stock by both opposite activation such as deterioration or amelioration.

3.4 Backlogging

A certain quantity of products required more than the available stock, resulting in a customer demand that was not met. It causes a product shortage, and the cost associated with this shortage is known as the shortage cost.

Excess demand is sometimes met without causing inventory to be depleted before the next replenishment. Backlogging is the term for when a demand is met but not yet fulfilled. To prevent losing goodwill customers, the shortage is sometimes partially filled with higher administrative costs. This is referred to as partial backlogging.

3.5 Order cycle

The time period between placements of two successive orders is referred to as an order cycle. There are two types of inventory review systems.

3.6 Continuous review

In this system, the inventory levels are viewed continuously until it reaches a specified point where new order is placed.

3.7 Periodic review

In this system, the inventory levels are viewed at equal time intervals and orders are placed at such intervals.

3.8 Time horizon

The time period over which the inventory level will be controlled the time horizon.

3.9 Inventory models

There are two types of inventory models: deterministic models and probabilistic models.

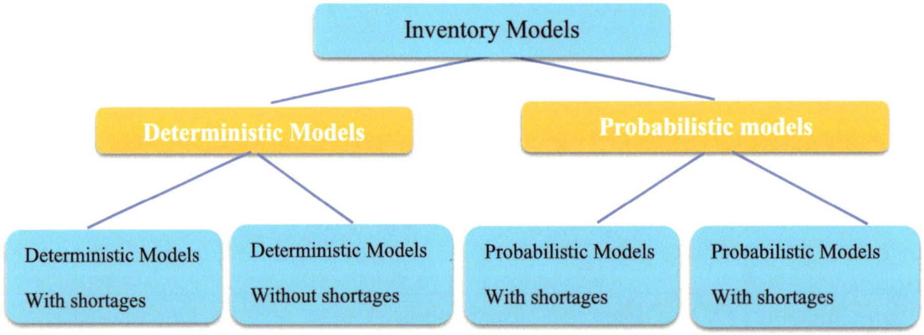

Based on business nature, inventory can be classified into two ways. One is production inventory and another one is trading inventory.

3.10 Trading inventory

The trader holds the trading goods inventory in organizations.

3.11 Production inventory

Production industry focuses on stock valuation, stock management, and stock control. Production inventory used raw materials in production, the semifinished goods in the warehouse, and finished goods are ready to sell. The production system produces different types of products, and it is supplied into different warehouses. This system maintained a sufficient raw material in storage for smooth running of an enterprise. This inventory needs to take practice for keeping enough stock on hand so the production process can run effectively.

Processing of each product type is carried out in lots of units, and finished lots are placed in the warehouse. Lot processing time is deterministic. Failures may occur during the production process when the production facility is busy. Times between failures are exponentially distributed with average units.

Production inventory is further classified into three inventories. There are raw materials, work-in process, and finished goods.

3.12 Raw materials

Raw materials can be commodities or components that are used in the production of goods and finished goods inventory. It can be classified as either direct materials or indirect materials. Material cost adds in production with the manufacturing overhead to calculate the raw materials inventory.

For example, automobile companies purchase various kinds of metals such as steel, aluminum, resins, copper, lead, and platinum from various suppliers that are used to make finished goods. i.e., cars. Flour is considered as indirect materials in bakery company to make finished products such as bread, pasta, crackers, many cakes, and many other foods are made using flour.

Raw materials include oil, lubricant, light bulbs, screws, nuts, bolts, etc., which are the examples of indirect raw materials.

3.13 Work-in-process

It is defined as the goods that are in different stages of production in production process. Products have been partially completed in the production cycle. Sometimes, products are waiting to be assembled and sold. Inventory level of work-in-progress should be kept as low as possible, which gives better returns to the industry. It can be avoided to lock up the capital of the enterprise.

3.14 Finished goods

Products are completely finished, which are ready for sale in the production process. That is, finished goods inventory is in stock that is available for customers to purchase. In production industry, the product is first made and then sold to the buyers or the received order is first, and then it is manufactured as per requirements with some specifications.

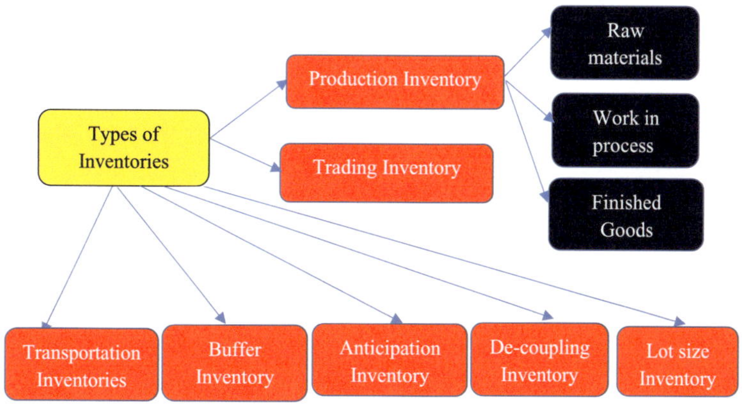

3.15 MRO

MRO represents the maintenance, repair, and operation supplies. These are materials, equipment, and supplies used in the production process at a manufacturing plant but are not part of the finished goods being produced.

3.16 Transportation inventories (pipeline inventories)

Transportation inventories are that inventory items are shipped from one distribution center to another, and significant amount of time is consumed in the transshipment of items.

3.17 Buffer inventory

It is an extra stored volume of products, which is kept to meet uncertainties of demand and supply. It is used to reduce the incidence or severity of stock-out situations in sales and thus provides better customer service.

3.18 Anticipation inventory

Anticipation inventory is an extra finished products or raw materials that a company must keep such inventory on hand to meet expecting a change in customer demand in the near future. This stock would provide benefits to companies by satisfying their customers if there is a surge in demand. These are built up in advance for the reason of large sales, a promotion program, or a plant shutdown period. It keeps men and machine ready for future needs. For example, keeping crackers well before Diwali or air coolers or air conditioners before summer.

3.19 Decoupling inventory

It is a type of inventory that is of use to store inventory. It is used to meet purchase orders in the case of inventory production slowing or stopping. If various production stages operate successively, then the breakdown of one or many may affect the entire system. This kind of interdependence is not only expensive but also disruptive for the entire system. The inventories used to reduce the interdependence of various stages of the production system are known as decoupling inventories.

3.20 Lot size inventory

It is the quantity of an item ordered for delivery on a specific date or manufactured in a single production run. That is, the amount of quantity produced/purchased during single production cycle. These inventories are stored in advance rather than the exact amount needed at a given time.

3.21 Backorder inventory models

A customer wants a new model two-wheeler bike Yezdi Roadster'22 model. It is not available in the stock in motor cycle show room. However, if the customer is willing to wait a few days or weeks, the retailer is able to order the bike. This type of shortage is called as backorder. In this backorder situation, the customer waits until the order arrives and then order is filled. The retailer should maintain a short period of waiting to receive the order and which is delivered immediately. The negative inventory level is occurred or incurred during the back order. It incurs a backorder cost in terms of the labor cost and special delivery cost while handling the backorders. In backorder situation, the retailer should maintain a goodwill customer or to avoid loss of goodwill because some customer will have to wait for their orders. If backorders can be tolerated by the customers, the backorder cost measures associated with the loss of goodwill when a customer wait for an order. See Ref. [3].

The items with high backordering costs will be handled with few backorders. Allowing the optimal number of planned backorders is projecting a minimum percentage savings in cost from the no shortage in inventory model.

3.22 Lowering inventory cost

As teamwork, coordination, and improvement of information sharing, opportunities are available for better cost control in the operation of inventory system. Some industries have a contract supplier under which the supplier regularly gives information regarding the status and schedule of upcoming production runs. The supplier's information concerning scheduled production runs provides a better understanding of the lead time involved for a product and the resulting lead-time demand. Information sharing by the supplier thus enables the order quantity, reorder point system to operate with a lower inventory holding cost.

Inventory may be checked and orders placed on a weekly, biweekly, monthly, or some other periodic basis. When a firm or business handles multiple products, the periodic review system offers the advantage of requiring that orders for several items be placed at the same present periodic review time. Some periodic review time with this type of inventory system, the shipping and receiving of orders for multiple products are easily coordinated.

3.23 Inventory management system

The industry bears a great deal of responsibility for producing high-quality goods by using appropriate natural resources and ensuring that they are delivered in good condition at the appropriate time. There are always some central issues such as preservation maintenance, product quality, and the need for good production in warehouse management. The right use of good products not only maintains good health but also makes it economically viable. It is more expensive to produce and distribute good-quality products than to produce low-quality items.

Some industries facilitate to use the advance technology for producing good quality of the products that stimulate the customers to buy quality products. This is because it does stick or seal on the product that provides the best-quality coating products. This technology would reduce product losses and increase the number of products available, ensuring the highest possible quality of products.

Many industries face a challenge in bringing good products to market with storage facilities using safety techniques to reduce perishability. Some technology is used to preserve liquid products and the industry invests in the cost/benefit of technology, which improves quality and reduces the rate of spoilage with good products.

Inventory management is affected by many factors such as unwanted defective items through processing, large quantities of stored products, improper distribution to distribution centers. Inventory requires more attention to allocating the right space to store products, preserving maintained goods, and fluctuations in the timely delivery of products. By holding an eye on inventory, the decision-makers can maintain a list of stock and what products are monitoring credits and remain in the warehouse. Because of the complexities of goods stored in warehouse with external technology, inventory systems must pay more attention to reduce the emission and power consumption in the processing industry. Inventory systems require accurate information about the inventory of goods to avoid the loss of inventory, creating more than profits.

In more general terms, the focus of the just-in-time philosophy is on *avoiding waste* wherever it might occur in the production process. One form of waste is unnecessary inventory. Others are unnecessarily large setup costs, unnecessarily long lead times, production facilities that are not operational when they are needed and defective items. Minimizing these forms of waste is a key component of superior inventory management.

3.24 Carbon emission reduction

Carbon taxes and cap-and-trade policies are being used by governments to reduce carbon emissions. The cost of carbon dioxide emissions is the subject of carbon tax policy.

A cap-and-trade policy is a type of trade policy that imposes limitations on carbon dioxide emissions reductions. Arash Sepehri [4] developed a model for controlling carbon emissions that included a carbon cap-and-tax policy in green technologies. In terms of productivity and social welfare, Xiaoping Xu proposed grandfather-based and benchmark-based allocation methods that will result in lower carbon emissions. See Ref. [5].

3.25 Block chain technology

Nowadays, block chain technology is the backbone of many organizations. Block chain is a cloud storage network, which shares out-of-records among members in the network space in a secure way. It is accountable to all and provides strong technical support for record transactions that ensure transparency for all members within the network. Block chain technology can increase trust and security by strengthening and collaborating on shared data knowledge about inventory in a centralized network based on various authorities and responsibilities.

Because inventory involves the difficulty of accurately estimating sales levels, production schedules, demand and consumption requirements, management needs block chain technology to make inventory decisions.

Block chain can greatly improve to support right decisions, the transparency of operations in any industry that becomes the heart of business strategy. It is used to maintain inventory data with either RFID tags or electronic product cods, which are unique identifiers or digital signatures. Transaction of any inventory data is recorded on the block chain, which transferred among involving parties who have rights to see the entire data that provides a complete, trustworthy.

Block chain technology is digital technology that is a decentralized ledger for recording inventory data transactions between multiple partners. Partners can view any variation in stock through this technology such as excess inventory or low stock inventory. Loss or fraud in the recorded data can be avoided. In the Qiu-xiang Li supply chain model [6], blockchain technology is considered to have an impact on the game and market demand between members of the supply chain through information sharing.

DOI: http://dx.doi.org/10.5772/intechopen.104387

Inventory control is an important part in logistics, which could manage the inventory by making revenue to keep continuous moving on to strengthen logistics and inventory management activities. Due to the need to retain this physical stock of goods in future usage, inventory management is one of the most important aspects of logistics supply management. The desired results and ensure measurable targets are set to drive the right decisions across the organization.

Author details

M. Rameswari
Department of Mathematics, SSM Institute of Engineering and Technology, Dindigul, India

*Address all correspondence to: sudharameswari1977@gmail.com

References

[1] Taha HA. Operations Research–An Introduction. Prentice Hall of India; 2001. p. 439

[2] Swarup K, Gupta PK, Mohan M. Operations Research. Sultan Chand & Sons; 2003. p. 365

[3] Anderson DR, Sweeney DJ, Williams TA. An Introduction to Management Science–Quantitative Approaches to Decision Making. Cengage Learning; 2010. p. 510

[4] Sepehri A. Controllable carbon emissions in an inventory model for perishable items under trade credit policy for credit-risk customers. Carbon Capture Science & Technology. 2021;**1**:100004

[5] Xiaoping X, Zhang M, Chen L, Yugang Y. The region-cap allocation and delivery time decision in the marketplace mode under the cap-and-trade regulation. International Journal of Production Economics. 2022;**247**:108407. DOI: 10.1016/j.ijpe.2021.108407

[6] Li Q-x, Ji H-m, Huang Y-m. The information leakage strategies of the supply chain under the block chain technology introduction. Omega. 2022;**110**:102616

Chapter 6

On Maritime Digitalization in Emerging Environments

Sanja Bauk

Abstract

The purpose of this chapter is to propose a conceptual framework for implementation of advance info-communication technology and systems (ICT&S) across maritime cluster in emerging environments, with emphasis on some South-East European countries and South Africa. Smart implementation and adoption of the ICT&S in shipping and port management have been considered due to the Holtham's & Courtney's model (2010). Furthermore, particular attention is given to Blockchain-based Maritime Supply Chain System (BMSCS) conceptual model, which includes distributed relational database, smart contracts, and crypto-currency payment mechanism as main constructs. The document workflow management, financial processes, and device connectivity are considered as key scenarios in the blockchain model. The stakeholders, who play the role of traditional intermediaries in the goods and accompanied documents and data flows, including overall supervision of the processes between end nodes, that is, shippers and customers, are encompassed within the BMSCS scheme. Some strengthens and frailty of BMSCS are highlighted, along with the suggestions for further investigation in this field, with the aim of better understanding prospective benefits and challenges of blockchain more extensive deployment across maritime sector in the future, with emphasize on developing environments.

Keywords: maritime, digitalization, emerging economies, conceptual framework

1. Introduction

Maritime is an important industry. Over 80% of the total transport of all goods takes place by sea because it is the most economical and most massive mode of transport. The world's seas provide free waterways. These are the largest absorbers of carbon dioxide and the largest producers of oxygen. The seas are the main source of food for one-third of the world's population. Oil and diamonds, e.g., are extracted from the seabed. However, the seas are exposed to pollution caused by both natural disasters and human factors. Another paradox related to the seas is that the maritime industry lags significantly behind other industries in terms of digitalization. Some facts that support this statement are as follows: a large number of ships do not comply with the requirements of the Safety of Life at Sea (SOLAS) Convention. Some ships do not have modern electronic navigation aids such as Electronic Chart Display and Information System (ECDIS) nor (Satellite) Automatic Identification System ((S)

AIS), for instance. Analyzes of accidents at sea have shown that the crew sometimes is not familiar with these devices. Digitization at land is more developed than at sea, and the main reason for this is the lack of profound investigation of Internet connectivity at sea, which can be hampered by sea surface movements, wave occlusions, rough weather, poor coverage, etc. Consequently, inter-organizational information systems (IOS) are used 75% in hinterland and only 25% in maritime. Internet of Things (IoT) is used considerably less at sea than at land. In road and rail transport, it is possible to track cargo at the level of a single unit or a container, while in maritime transport this is still not possible. As an example, we can use transport of dangerous goods. The casks (drums) with radioactive waste (plutonium, e.g.) can be tracked by Radio Frequency Identification (RFID) chips, Internet connection, and security backend web applications at the level of a single freight unit or a drum in the road and in rail transportation [1, 2], but not in maritime [3]. Furthermore, there are a large number of autonomous vehicles on roads (about 1.5 thousand) and in the air, i.e., drones (about 1.5 million), but only one autonomous ship (Yara), and another one is currently under construction [4]. Blockchain technology is not widely accepted yet, since there are various impediments like the lack of trust between stakeholders; government support; legislation; standards; along with the stakeholders' readiness for risky investments in emerging technology. Some extensive desktop studies of academic writings shown that a very small percentage of articles deal with advanced info-communication concepts such as big data, virtual intelligence, robotics, 3D, virtual reality, digital security, etc., in maritime. There is no clear political strategy for further development of info-communication systems in maritime. This complicates maritime digitization in developed countries, and considerably more in developing ones. Concerning the latest, this chapter is organized in the following manner: Section 2 deals with smart adoption of advanced ICT&S in general, and in maritime business. Additionally, a case study has been conducted in several non-EU and EU countries in this respect, based on the Holtham's & Courtney's model. Section 3 considers rational blockchain adoption in maritime, with a focus on developing environments, concerning BMSCS, TradeLens, smart contracts, and Blockshipping. This section also encompasses a case study on blockchain smart adoption in maritime business in emerging economies, with emphasis on South Africa and Montenegro, while Section 4 gives some concluding remarks.

2. Smart adoption of ICT&S in maritime

We live in a time of massive progress in the field of info-communication technology and systems (ICT&S). The question is do we really need all these innovations and do they always make our lives easier and better. In order to get the best out of ICT&S, we need to know which of these technologies we actually need and how to use them purposely. When it comes to a business environment, it is very important that higher management structures are aware of these needs and discuss them with employees. This is especially important in maritime business, bearing in mind that, stakeholders in maritime are usually conservative and not early new technology adopters. Several studies considered ICT&S rational, intelligent or smart implementation and adoption of advanced ICT&S in maritime [5–10]. On the Holtham's & Courtney's model, in this chapter, it has been examined how stakeholders in maritime assess key constructs in the model, regarding some developed and developing countries in Europe. The considered countries were treated as the European Union (EU) and non-European Union (non-EU) ones. Before presenting the methodology and the obtained results, the overview of the applied Holtham's & Courtney's model is given.

2.1 The Holtham's & Courtney's model

This model is composed of four constructs: knowledge, ICT&S strategy, system effectiveness, and ICT&S management. These constructs are supported by organizational culture and top manager's mindset (**Figure 1**).

Knowledge can be described as the understanding of a subject that one gets by experience or study, known either by one person or by people generally; or, as the state of knowing about or being familiar with something [11]. Furthermore, Cambridge dictionary depictures knowledge as awareness, understanding, or information that has been obtained by experience or study, and that either is in a person's mind or possessed by people generally; or, as skill in, understanding of, or information about something, which a person gets by experience or study. Symbolically, knowledge is one of the steps in the so-called ladder of knowledge. This ladder of knowledge encompasses data, information, knowledge itself, and wisdom stairs. The data and information have easier explanations than the concepts of knowledge and wisdom. In this context, the focus will be on knowledge stair, as a key for understanding contemporary ICT&S and their rational implementation in maritime business. Knowledge here means consciousness about advanced ICT&S availability at the market, including the ICT&C purposiveness regarding particular business strategies, processes, and activities.

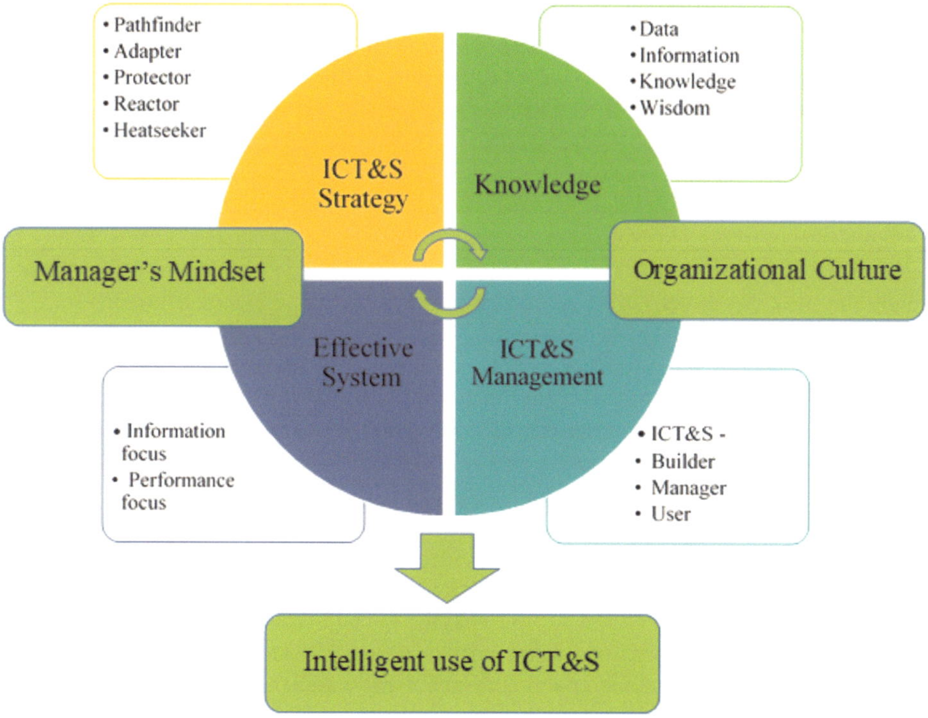

Figure 1.
Adapted Holtham's & Courtney's model (source: Own).

The ICT&S strategy brings together business and technology. Due to the Holtham's & Courtney's model, there are five different strategic orientations.

Pathfinder: Systematically seeks and selectively exploits relevant ICT&S trends to gain competitive advantage and enable entry into new markets. Pathfinder is willing to experiment with new, advanced ICT&S. Pathfinder constantly seeks a competitive advantage by detecting or sensing emerging ICT&S trends and opportunities.

Adapter: Functions at two types of market: one relatively stable and focused on efficiency, and the other where ICT&S plays an increasingly important role. Adapter applies different rates of technological uptake in each. This split feature is typical for businesses where different rates of technological adoption are present, and adaptability rather than uniform solutions are applicable.

Protector: Meticulously assesses ICT&S investment for its efficiency orientation and applies technology primarily to reduce costs of investments and increase communication processes rather than new market opportunities creation. Protector is control orientated and slow to innovate. These organizations work in domains where core ICT&S-based technologies are commonly available and easily replicable.

Reactor: This is a characteristic of an organization where technology is not perceived as a strategic tool. It responds slowly to change and tends to view ICT&S applications as standalone tools. In this strategic orientation, the ICT&S platforms usually appear to be weak or obsolescent. The risk is that the organization could quickly become non-competitive through a lack of capacity to meet customers' needs.

Heatseeker: Seeks upon ICT&S fashioned instead of strategically analyzing the best ICT&S fit for its business processes. Heatseeker is typical for an organization whose structure is in constant change, experimenting with innovations before obtaining steady business performance. This organization is receptive to ICT&S spending and subsequent partial reversals when intended benefits are not realized quickly.

The system effectiveness can be achieved by setting and communicating critical success factors (CSF) [12] and developing them steadily. The first step is to use technology to create an effective operational platform, primarily with internal information. Then, the CSFs can be widened to foster improved skills to use technology. This will start with employees and then extend to suppliers and customers. Once when these two steps work well, the CSFs can be broadened to encompass external information about markets, customers, and competitors. After these, three steps comes business intelligence, which allows organizations to identify and manage risk while developing new products, services, and markets to ensure a successful future.

The ICT&S management is based on ICT&S builders, ICT&S managers, and ICT&S users. A person or management team that communicates the needs of ICT&S users into ICT&S builders or designers has to be engaged in the organization as a knowledge navigator, or information resource manager. There are business organizations, which recognized this triangle and which are working on filling and improving the personnel skills towards achieving this goal [13, 14].

These four constructs, which form the backbone of the Holtham's & Courtney's model are underpinned by organizational culture and top manager's mindset.

Concerning the organizational culture, there is universal agreement that it exists and plays an important role in shaping behavior in organizations. However, there is little consensus on what organizational culture actually is. Here are quoted several expressions that can be used in absence of universally accepted one [15]: organizational culture is how organizations do things; organizational culture is the sum of values and rituals, which serve as glue to integrate the members of the organization; organizational culture is civilization in the workplace, etc.

Top manager or top management team's role is to weave a fabric of horizontal (information, technology, people, and organization) and vertical (direction, knowledge, process, and climate) threads mutually intertwined. In organizations where knowledge is a core dimension, managers have frequently identified people's skills as the major influence, along with organizational climate. Moving from the information-based to the knowledge-based enterprise is a major challenge for today's companies [16]. Therefore, managers have to combine proper notions from several different domains: organizational behavior, human resource management, big data, analytics, artificial intelligence, etc. Technology is a key enabler, but not usually as significant as skills and climate. Top managers' team mindset covers all considered constructs and it affects intelligent or rational use of ICT&S.

These constructs are used in the following analysis as independent variables. As a control variable is used non-compliance between technology-led potential and its everyday usage, while the dependent variable is intelligent use of ICT&S, which reflexes efficient and smooth communication between tasks, technologies, and employees [4].

2.2 Case study on smart ICT&S adoption

Based on the Holtham's & Courtney's model, a survey was conducted among stakeholders in maritime in terms of how rationally they use the ICT&C in their business. Selected stakeholders from four EU (Croatia, Greece, Italy, and Slovenia) and from four non-EU countries (Albania, Bosnia & Herzegovina (B&H), Montenegro, and Serbia) were included in the survey. Forty experts (five per each considered country) from maritime administrative bodies, agencies, private marine companies, ports, and universities (maritime departments) were asked to evaluate the set of fifteen statements by means of Likert (1–5) scale (**Table 1**). The last two statements refer to the control and dependent variables in the model, respectively, while the rest of the statements correspond to the independent variables.

The level of respondents' agreement with the proposed statements is labeled due to the following scheme: (i) if the average score per group of experts from a certain country is between 1 and 2.5 then the level of experts' agreement is "low"; (ii) if it is between 2.6 and 3.5, then the level of agreement is "moderate", and (iii) if it is between 3.6 and 5, then it is "high". The results of the survey are presented in **Table 2**.

According to the results presented in **Table 2**, it is obvious that all respondents evaluate knowledge, ICT&S management, and manager's mindset as highly important for rational application of ICT&S in maritime organizations. When it comes to the system effectiveness and organizational culture, the experts assessed these constructs as high or moderate important for intelligent implementation of ICT&S. Control variable confirms validity of the assessments assigned to the dependent variable "intelligent use of ICT&S in maritime". Namely, all experts highly agreed with the statement "the ICT&S serve as a connective tissue among tasks, technologies, and employees in your organization", which in fact justify smart exploitation of ICT&S in maritime business. When it comes to ICT&S strategy orientation, the respondents are dominantly *adapters*, while only one of the respondents is *reactor*. The *adaptors* are looking for the third path, while reactors leg behind regarding adopting new ICT&S. Adaptation means that emphasis is on modification rather than fundamental reconfiguring of the existing ICT&S. On the other side, reaction means using weak and obsolescent ICT&S platforms. The reasons behind this orientation should be examined through in-depth interviews with respondents and through further analysis

Construct	Statements
C1: Knowledge	S1.1: Knowledge is important for business success. S1.2: Knowledge and skills of employees are important for efficient and effective use of ICT&S?
C2.1: ICT&S Strategy *Pathfinder*	S2.1: New ICT&S solutions adoption is risky for the organization.
C2.2: ICT&S Strategy *Adapter*	S2.2: Analyzing carefully the existing ICT&S solutions prior to their introduction into the organization is important.
C2.3: ICT&S Strategy *Protector*	S2.3: The ICT&S reduces operational costs of the organization.
C2.4: ICT&S Strategy *Reactor*	S2.4: The available ICT&S solutions can be adapted to the current business needs of your organization.
C2.5: ICT&S Strategy *Heatseeker*	S2.5: The latest ICT&S solutions are the best ones.
C3: System effectiveness	S3.1: Your customers intensively use ICT&S resources of your organization (web site and various online users' apps). S3.2: The ICT&S allows you to become familiar with the current market trends in the area of your business.
C4: ICT&S management	S4.1: The ICT&S functions are important for the successful functioning of the organization and its business success. S4.2: The usage of ICT&S for operational takes within your organization (accounting operations, database of employees, database of business partners, market analysis, etc.) is extensive.
C5: Culture	S5: Positive organizational culture and clime are important for effective use of ICT&S.
C6: Manager's mindset	S6: Manager's mindset is important for effective use of ICT&S.
C7: ICT&S capacities versus exploitation	S7: There is a divergence between ICT&S capacities and their real application on a daily basis in your organization.
C8: ICT&S intelligent exploitation	S8: The ICT&S serves as connective tissue among tasks, technologies, and employees in your organization.

Source: Own.

Table 1.
The survey content.

of their business strategies, including the position of the ICT&S in it. This might be the subject of further investigation in the field.

In addition to the analysis of the degree to which ICT&S are used rationally in the maritime business, it was also examined which advanced info-communication platforms are available to the maritime organizations in which the respondents in this study work. The results of this part of the survey are presented in **Table 3**. It is evident that there are efforts to modernize ICT&S in maritime business, but also that some of the analyzed maritime organizations, i.e., countries are lagging behind, especially those that are not members of the EU. Namely, the non-EU countries have to reconsider their business development strategies and ensure funds for implementing new ICT&S and renewal of the existing ones. These countries should follow actual trajectories and scenarios towards efficient and effective digitalization in maritime [17].

Through this case study, it is shown that responders, who are employed in maritime administration and business organizations in four EU and four non-EU countries, have similar attitudes towards concerned constructs inherent to intelligent exploitation of contemporary ICT&S. They all identified knowledge, ICT&S management,

Constructs		C1	C2	C3	C4	C5	C6	C7	C8
	Country	Knowledge	ICT&s Strategy	System effectiveness	ICT&S management	Organizational culture	Manager's mindset	Control variable	Dependent variable
EU	Croatia	High	Adapter	High	High	Moderate	High	Low	High
	Greece	High	Adapter	Moderate	High	Moderate	High	Moderate	High
	Italy	High	Adapter	High	High	High	High	High	Low
	Slovenia	High	Adapter	High	High	High	High	High	Low
Non-EU	Albania	High	Adapter	Moderate	High	High	High	High	Low
	B&H	High	Adapter	High	High	Moderate	High	Moderate	High
	Montenegro	High	Adapter	Moderate	High	Moderate	High	Moderate	High
	Serbia	High	Reactor	High	High	Moderate	High	Moderate	High

Legend: *High agreement; Moderate agreement; Low agreement*
Source: Own.

Table 2.
The assessments of the constructs.

system efficiency, organizational culture and manager's mindset as key perpetuators of rational and purposeful use of the ICT&S. This speaks in favor of their sound education and awareness about the importance of ICT&S in today's dynamic maritime business environment. Regarding ICT&S strategical orientation, the respondents are cautious, i.e., not prone to take risks of investing in new ICT&S solutions and experimenting in the market.

When it comes to the availability of common and advanced ICT&S in analyzed maritime organizations, it is shown that there are big differences among EU and non-EU countries. For instance, maritime organizations in Slovenia have almost all listed ICT&S except blockchain technology, AGVs, digital twins, are UxVs. The companies in Italy have, e.g., digital twins and UxVs. Croatia and Greece have also quite an extensive list of available ICT&S. On another side, explored non-EU countries are modestly equipped. The examination of the reasons for such difference and how it can be alleviated in order to avoid disruptions in maritime ecosystem and negative economic implications for the non-EU countries should be the subject of further, more profound studies. None of the considered maritime entities does have on disposal Blockchain-based Maritime Supply Chain System (BMSCS), e.g. Since BMSCS is the advanced emerging ICT&S platform in contemporary maritime business, the following text attempts to explain the basic principles of this platform, including its benefits and challenges.

3. The BMSCS conceptual framework

Bockchain in maritime is a far broader system than cryptocurrency-based electronic financial transactions mechanism. In literature, it is named as Blockchain-based

Available ICT&S	Countries							
	Non-EU countries				EU countries			
	Albania	B&H	Montenegro	Serbia	Croatia	Greece	Italy	Slovenia
Electronic Data Interchange (EDI)		×	×		×	×	×	×
Enterprise Resource Planning (ERP)		×	×		×	×	×	×
Customer Relationship Management (CRM) System	×	×	×		×	×	×	×
Electronic Logistics Marketplace (ELM)		×				×	×	×
THETIS (PSC—Port State Control)			×		×	×	×	×
Blockchain-based Maritime Supply Chain System (BMSCS)	▼	▼	▼	▼	▼	▼	▼	▼
Automatic Identification System (AIS)	×	×	×	×	×		×	×
Long-range and tracking (LIRT)	×		×		×	×		×
Vessel Traffic Monitoring Information System (VTMIS)	×		×		×	×		
Sea Traffic Management (STM)	x				×		×	×
e-Navigation					×			×
e-Maritime					×			×
Common Maritime Communication Platform (CMSP)					×			×
Maritime Surveillance Services (MSS)	×				×			×
SafeSeaNet (SSN)			×		×	×		×
Maritime Single Window (MSW)					×	×	×	×
Automatic Guided Vehicles (AGV)								
Digital twins							×	
Remotely controlled vessels	▼	▼	▼	▼	▼	▼	▼	▼
Unmanned sea or underwater vessels (UxVs)						×	×	
Earth Observation Services —Search and Rescue (SAR) sensors					×	×	×	×

Available ICT&S	Countries							
	Non-EU countries				EU countries			
	Albania	B&H	Montenegro	Serbia	Croatia	Greece	Italy	Slovenia
Earth Observation Services —Optical sensors					×	×	×	×
Satellite-based oil spill detection system at sea			×		×			×
Oil spill prediction modeling system			×		×		×	×

Legend: *There is no ICT&S of such type*
Source: Own.

Table 3.
Available ICT&S in the examined maritime organizations.

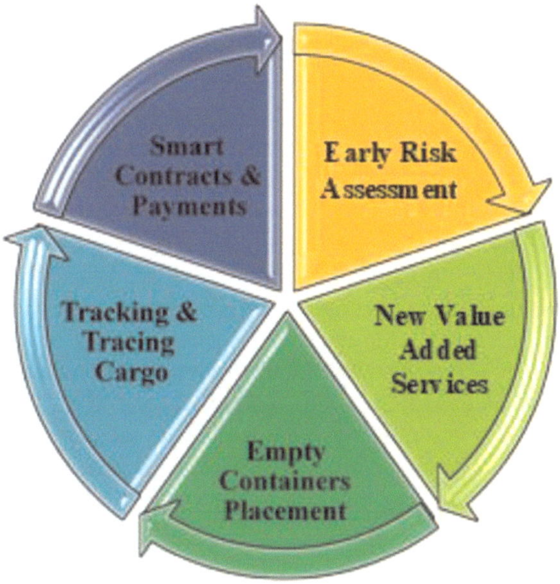

Figure 2.
BMSCS key components (source: Own).

Maritime Supply Chain System (BMSCS) [18]. It includes smart contracts and payments, tracking and tracing cargo, empty containers placement, early risk assessment, and services that can create new added values in maritime (**Figure 2**).

Maritime is an indispensable link in global supply chains. Hereof, blockchain technology is likely to become unavoidable in shipping and port management, in striving to optimize global supply chains and make these more efficient and effective. The BMSCS should reduce the volume of administrative work, errors that occur due to manual work, delays in the delivery of goods, and consequently overall costs of transportation and delivery of goods. Due to some studies, the costs of global supply chains should be reduced by approximately 15% by implementing blockchain technology [19].

In addition to these advantages, blockchain also has certain disadvantages. Maritime sector is generally risk-averse, tending not to be an early adapter of new potentially risky technology [20]. Some stakeholders in maritime want to keep their data secret, since "competition is fierce" and "a lot of industry actors are basically competing with the same service" [21]. In other words, some partners in the supply chain consider information as a competitive resource and are unwilling to share them. Positional data might be used to track vessels by identifying port locations, fueling locations, and routes [22]. This is particularly the case with tracking dangerous and hazardous goods, pharmaceuticals, or food. The use of blockchain does not guarantee that the information recorded in ledgers is correct and does not prevent tampering data prior to entering it into blockchain ledger, e.g., the contents of a container, fuel production, testing or combustion, and the like [23]. Due to the huge amount of data and traffic generation, including data storage, blockchain requires a wideband like G5 or G6 [24], while the internet speed can be low when the working stage is offshore. Further, blockchain causes high-energy consumption [25]. Blockchain in the maritime sector indicates the potential to reduce transaction costs in a number of areas, including reducing the need for intermediaries such as brokers and courier services and reducing related financial expenses and energy costs. However, one should not forget that this does not include the costs of the overall investment and expenses associated with blockchain implementation and adoption, especially in developing or emerging economies [26, 27].

The present level of awareness, knowledge, and expertise about blockchain is scarce among the stakeholders. Therefore, educational, training, or capacity-building programs are necessary at regulatory, administrative, and operational levels. Higher level of standardization across the global supply chain is necessary as well. The Digital Container Shipping Association (DCSA) conducts efforts in this regard, but further, more extensive, actions are necessary. In general, there is a hesitation by stakeholders in maritime sector to invest in blockchain systems in terms of technological integration, regulatory, organizational, and educational costs, since the maritime sector traditionally relies on its legacy systems. There appears to be a gap between what practitioners in the blockchain area suggest and what has been a range of state-of-the-art approaches in software engineering and information security research and practice [28]. Furthermore, the major liner shipping companies are the most likely parties to benefit from blockchain regarding the complexity of their blockchains and huge requirements on financial resources [29]. This can put other potential actors in the global supply chain at a disadvantage. The last but not the least, the basic attitude should be that technology, in this case, blockchain on the top of the global supply chain should improve the human condition, and not replace humans [30]. Therefore, human and ethical dimensions of blockchain technological development and more extensive deployment, should not be neglected.

In the following text, we shall present blockchain framework in maritime at the example of two applications: TradeLens and Blockshipping. TradeLens is used for tracking and tracing cargo along the global supply chain, early risk assessment, smart contracts, and value-added services created through an open platform. Blockshipping is used for empty containers' optimal placement by autonomous intelligent software agents.

3.1 TradeLens

TradeLens is a new business model in shipping and port management. It enables one-to-many connections for all the actors, all individuals that are involved in a global supply chain instead of bilateral connections. Everybody come together in a maritime industry-neutral, open platform for every participant [31]. Maersk, the world's largest international container shipping and logistics company, and IBM the technology leader in blockchain came together to provide a new, open platform solution underpinned by blockchain to help unlock some of the opportunities for a more efficient global supply chain. Maersk and IBM have a long history of working together, actually decades. In March 2017 these organizations collectively try to improve global trade through digitization. In January 2018, they launched an early adapter program; trials began, and in August 2018, they formally launched the TradeLens limited availability platform, shared among 92 participants. In December 2018, TradeLens is commercially realized, along with 1.5 million events per day published to the platform. Some types of these events are presented in **Table 4**.

The platform can track 120+ unique consignment shipments, while 60+ network members are onboard or in a process of accessing. TradeLens supports 18+ unique, standardized, trade document types. Some of these documents are listed in **Table 5**. In February 2019, enhanced document sharing, permissions, and notifications were released. The platform includes half a billion events on annual basis and this number grows with more and more network members.

More than twenty million containers of cargo information are in the system today, which is roughly 1/5 of global trade and it is growing. The platform involves numerous parties and systems: ocean carriers, ports and terminal operators, inland carriers, shippers, consignees, beneficiary cargo owners, freight forwarders, 3PLs, custom authorities, government agencies, financial and insurance services, transportation management systems, Port Community Systems (PCSs), supply chain validity systems, supply chain, manufacturers, retailers, etc. They all collaborate and share information. TradeLens provides them with comprehensive, real-time visibility and immutability across the end-to-end journey of shipment. In other words, data is available immediately, along with the single simplified view across all shipments.

Actual	Estimated	Planned
Start container tracking	Documentation cutoff: Vessel ETD	Import documents approval
Start shipment tracking	VGM (Verified Gross Mass): Vessel ETD	Discharged from truck
Booking confirmation	Cargo cutoff: Vessel ETD	Loaded on vessel
Stuffing started	Rail ETD	Stuffing started
Vessel ATA	Rail ETA	Stuffing completed
Vessel ATD	Bill of Lading Available	Loading on vessel
Loaded on rail	Vessel ETD	Gate in
Rail ATD	Vessel ETA	Gate out
Rail ATA	Discharged from vessel	Packed container selected for inspection
Loaded on truck	Load on vessel	Packed container passed inspection
+Add more	Custom release	Cargo specific certificate approved
	+Add more	+Add more

Source: [31].

Table 4.
TradeLens standardized events.

Document	Party
Import documentation approved	Customs House Broker
Customs release	Customs Authority
Cargo geography-specific certificate approved	Customs House Broker
Bill of lading available	Beneficiary Cargo Owner (BCO)
Certificate of origin available	Beneficiary Cargo Owner (BCO)
Packaging list available	Beneficiary Cargo Owner (BCO)
Commercial invoice available	Beneficiary Cargo Owner (BCO)

Source: [31].

Table 5.
TradeLens standardized documents.

For instance, a terminal operator publishes a piece of information about the fact that a container has been loaded onto a ship that becomes immediately available to everybody else in the supply chain. The idea is to build workflow based on smart contracts using chain code to derive cross-organizational workflow by excluding manual work.

Blockchain on which the platform is based, enables the trust in data that are available on the platform. It is an open and censorship-resistant distributed database model, secured by encryption and decentralization. Blockchain records information in blocks on a shared ledger, storing a synchronized copy of it on all the systems participating in the network, hence assuring its immutability. The trust anchors, which are the blockchain nodes, ensure through consensus algorithms that the information should be written on the platform as approved like valid. All information are auditable, verifiable, and temper proof; so, as soon as a piece of data is published to the blockchain it cannot be edited. The only way to edit a document is to create a new version of the document. Consequently, all the documents are fully auditable. Additionally, cryptographic hash of the data is written to the blockchain, and this is a part of the supply chain. It is important to say that private data remain private. TradeLens as an information-sharing model allows ecosystem partners to have access to the information they should access and vice versa. The platform offers a high level of flexibility through application of RESTful APIs (Application Programming Interfaces), back-end ERP (Enterprise Resource Planning), and secured front/back-end web services.

In the middle or at the very core of TradeLens solution, there is the platform and blockchain behind it. Below the platform is the network. The network is not a physical network. It is a set of entities that provide the data, including the data itself. The ocean carriers, ports, terminal operators, customs, shippers, inland transporters, etc., provide the data. On the top, above the platform are applications and services, i.e., RESTful APIs, back-end ERP, and secured web that enable people to exchange information. These are based on open published industrial UN/CEFACT standards that are defined at the platform level, so that third parties are allowed to build new value-added services and applications. This is the basic kind of model, through which TradeLens is moving forward as a paradigm shift in information sharing across the whole ecosystem. A conceptual framework of TradeLens as a blockchain-based solution in the global supply chain is presented in **Figure 3**.

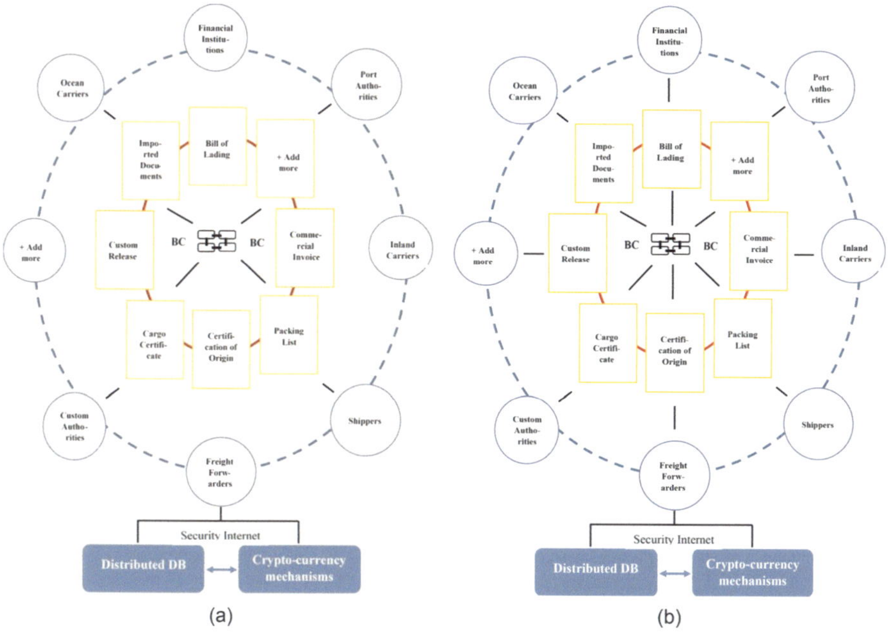

Figure 3.
TradeLens conceptual framework (source: Own).

What kind of information is shared across the platform, i.e., over the entire supply chain? This information is mostly shipping milestones. The information on: has a container been staffed; has the container be gated; what is the estimated time of arrival (ETA) of the container at the destination, and so on, are in fact shipping milestones. However, it is more than that. It is also the documents in maritime, both structured and unstructured (like PDFs, scans, images, etc.), by making them available to the participants along the supply chain. The documents need to change 'hands'. They need to be approved, updated, and available to build workflow using smart contracts, like the bill of lading, clearance, insurance, etc. This is powerful in terms of driving cross-organizational data flow in maritime.

Within TradeLens, there is sensor data and IoT for referring to the container number, electronic seal, temperature inside it, for instance, etc. All of that is part of the underlined data that is made available to the participants who need that data. There is a whole concept of seamless and permission data-sharing model that is built on the base: what your role is, i.e., are you terminal, ocean carrier, shipper, inland transporter, etc. The default permission model allows people to share information, so that information is made available to those who need it, but it is not available to those who should not see it.

TradeLens is of utmost importance whenever planned actions turn into unplanned. For instance, the ocean carriers' decision has implications not just for them but for all stakeholders further down the supply chain from customs brokers, port authorities, and terminal operators to inland transporters and consignees. With TradeLens, changes to the shipment are reflected immediately allowing supply chain participants to coordinate actions tightly, delivering the consignee's inventory in time. TradeLens allows near-instant logistics adjustments so the disruptions are kept to a minimum. Global trade is an incredibly complex system, but TradeLens and blockchain create an industry-wide and innovative solution to alleviate this complexity and related impediments.

3.2 Smart contracts

Smart contracts are programs stored on a blockchain that run when predetermined conditions met. They are used to automate the execution of an agreement so that all participants can be immediately certain of the outcome, without an intermediary's involvement or time loss. They can also automate a workflow, triggering the next action when conditions are met [32]. Smart contracts work by following simple "if/when (...) then (...)" statements that are written into code on a blockchain. In the other words, a smart contract presents the lines of code that are stored on a blockchain and automatically executes when predetermined terms and conditions are fulfilled [33]. A network of computers executes the actions when predetermined conditions have been met and verified. These actions could include releasing goods, funds, or confirmations in maritime supply chain. The blockchain is updated when the transaction is completed. That means the transaction cannot be changed, and only parties who have been granted permission can see the results. Blockchain network controls access. Within a smart contract, there can be as many stipulations as needed to satisfy the participants, so that the task will be completed correctly. To establish the terms, participants must determine how transactions and their data are represented on the blockchain, agree on the "if/when (...) then (...)" rules that govern those transactions, explore all possible exceptions, and define a framework for resolving disputes. Then, the smart contract can be programmed by a developer, although organizations that use blockchain for business, provide templates, web interfaces, and other online tools to simplify structuring smart contracts. Key benefits of smart contracts are speed, efficiency, accuracy, trust, transparency, and security (blockchain transaction records are encrypted, which makes them very hard to hack; plus, each record is connected to the previous and subsequent records on a distributed ledger, and hackers would have to alter the entire chain to change a single record). In maritime supply chain, sea waybill or bill of lading can be converted into a smart contract, while it requires an agreement between shipper and carrier, and/or any other relevant and permissioned parties to view the consignment, transport equipment, and documents, as permissions allow [34]. The benefits of such smart contract include simplified transmission of shipping instructions; management of document status and versioning; faster submission of shipping instructions for creation of final bill of lading; quick sharing of documents with all permissioned parties; including immutability, traceability, and auditability of the documents involved.

3.3 Blockshipping

Today the container shipping industry accounts for around 60% of all the world seaborne trade. This valuable industry has been troubled for years by challenges like overcapacity, low freight rates, security threats, and increasing environmental regulations. Currently, there are about 27 million containers in the world, which are moved from one destination to another on trucks, container cars, ships, rail, or waiting in the port, container yard, railway station, and the like. About 5 million containers are uncontrolled and nobody knows their precise locations; if they are currently in transit or waiting for collection. Consequently, no one knows if they are empty or loaded, which means that no one knows if a truck or a train is wasting time and energy carrying an empty "metal box" instead of carrying goods [35]. This is a huge waste of energy; it produces additional costs and negatively affects the environment.

Therefore, the global shared container platform (GSCP) is currently under development. As the world's first blockchain-based container registry, it will allow the industry to help real-time track all containers worldwide. The platform will enable the industry players to manage efficiently all kinds of transactions related to container handling. The GSCP has several user groups like shipping lines, leasing companies, banks, financial institutions, blockchain container investment syndicates, transport service providers, beneficiary cargo owners (BCO), container terminals, container depots, repair shops, etc.

Through a secure login, each user group will have a unique set of functionalities that matches their exact needs. For example, if you are a shipping line export user, you can use GSCP platform to find street turn matching opportunities for ensuring that empty containers meet export demands. You will see an inventory list of all export bookings, which require an empty container to the customer location for stuffing, rather than transporting an empty container from the port or the depo. For convenience, the platform enables users to switch between list and map view. The user can apply one or more filters and inventory will update accordingly, for instance, only showing FEUs (40-feet units). Any set of applied filters can be saved in user's filter presets. This way they are quickly accessible whenever the user needs them. Matching export containers with import containers is easy and swift. This enables both importer and exporter to save an empty container haulage trip, plus gate in and gate out fees at the terminals. The system identifies possible matches based on container size, type, boarding date, previous commodities carried, and availability. The platform also enables sending a request to the involved shipping line with the comment. The GSCP provides various ways to import booking and container data. The user can use EDI and API connection with the in-house booking or order platforms [36].

Blockshipping is, in fact, a shared pool of containers, which enables a "just in time containers" situation. Today more than 40% of all containers in transport are empty. Therefore, resources are wasted and costs are increased. With Blockshipping saving potential for the shipping industry might be at least 5.7 billion USD and reduction of CO_2 emission can be 4.6 million tons yearly [37]. Blockshipping platform is a part of the so-called programmable economy. In such an economy, the interactions among different parties will not occur through mediation of a third trusted party, but automatically through autonomous intelligent software agents (AISA). These are also called dApps (distributed Apps) that run on blockchain and are authorized and instructed by the parties involved in the BMSCS to negotiate autonomously on their behalf.

Four key subsystems of Blockshipping are [38]:

- Global Shipping Container Registry (GSCR), that holds real-time information about every container available through Blockshipping;
- Empty Container Repository Engine (ECRE) that continuously calculates the next best-laden transport for each container. The engine also "understands" the position of every truck available to transport the empty container;
- Autonomous Intelligent Software Agents (AISA) that run on blockchain and negotiate all agreements;

- Smart Contracts (SC) that can be treated as rental contracts established through autonomous negotiations, which persist on the blockchain and govern the rental through binding self-enforcing rental agreements.

Blockshipping has developed a unique revenue-sharing model, while it issues two types of tokens:

- Internal utility token, or Container Platform Token (CPT), and
- External revenue share coin, or Container Crypto Coin (CCC).

The CPT will be used for clearing and settlement of transactions between the users of the platform. These transactions will relate to many different services and fees. A percentage of the revenue goes to a revenue share pool and is passed on to the owners of CCC tokens. Blockshipping exchanges the CPTs in the revenue share pool to Ether via USD. Then, Blockshipping uses smart contracts to convert revenue Dutch auction on the Ethereum blockchain in which Blockshipping offers the owners of the CCC tokens price for their tokens. The offered price will increase until all available Ethers are spent. After the auction, Blockshipping distributes the acquired CCC tokens to all the owners of CCC tokens on a pro-rata basis. In this way, token owners are rewarded regardless of their decision to sell or keep their CCC tokens [39].

The processes flow within Blockshipping is based on several simple and fully automated steps. The easiest way to make an explanation is to follow an example. Let us assume that the shipping line needs to rent a container to transport goods from Nairobi (Kenya) to Rotterdam (Netherlands). Blockshipping empty container repository engine identifies the best-positioned empty container in Nairobi and informs the shipping line about the options. The shipping line informs its autonomous intelligent software agents (AISAs) about the containers. The rental negotiations then happen unsupervised between the shipping line and the container owner through their autonomous agents. The agreements established by AISAs are persisted on blockchain in smart contracts that govern the rental in binding self-enforcing rental agreements. Blockshipping container platform tokens CPT are used to pay rental fees, while the fees are transferred from the shipping line wallet, in accordance with the smart contract and reserved payment. Smart contracts can be changed if conditions change. For example, if the rental period is extended when the container reaches its final destination in Rotterdam. Then, the smart Oracle blockchain enforces the smart contract. The rental ends and releases CPTs to the container owner's wallet [40].

In addition to TradeLens and Blockshipping, there are a number of other blockchain applications in maritime. Some of these are given in **Table 6**. Concerning safety issues, semi-private blockchains are common. The consortium companies' reputation speaks, in fact, about safety. Maritime, as conservative, assesses and recognizes the quality of operation in long run. As we said previously, stakeholders in maritime are not early adopters. However, trust between network participants is a bigger problem than safety. The blockchain is an unorthodox technology and cryptocurrencies are still highly volatile. In such a setting, maritime stakeholders do not like to disclose essential business information about customers, suppliers, and cargo. Many freight forwarders and intermediaries, e.g., earn their profit thanks to information asymmetry. Interoperability will be a smaller problem in terms of technology (since standards have been developing) than in terms of trust and smooth process flows at (inter-)organizational level.

No.	Blockchain		
	Consortium	**Platform**	**Ledger**
1.	Port of Koper, Slovenia	CargoX	Public
2.	Malaysia's West Port & LPR - Brazilian textile importer	300cubits	Public
3.	Maersk & IBM	TradeLens	Consortium (permissioned)
4.	Abu Dhabi Ports and Port of Antwerp	Silsal	Consortium (permissioned)
5.	EY & Guardtime	Marine Insurance Blockchain	Public
6.	PIL, PSA & IBM	Proof of Concept (POC)	Consortium (permissioned)
7.	Port of Antwerp with Belfruco, Enzafruit, PortApp, 1-Stop and T&G Global	Smart Contracts	Consortium (permissioned)
8.	2021.AI Den Danske Maritime Fond, EUDP, INVICTA	Blockshipping	Public
9.	Port of Malmo & Port of Copenhagen	PortChain	Consortium (permissioned)
10.	AAT, FileVersion Health, CROP	CargoChain	Consortium, (permissioned)

Source: Adapted from [41].

Table 6.
Some blockchain applications in maritime.

3.4 Case study on blockchain adoption

Within this case study, we explored how maritime stakeholders in two developing countries, South Africa and Montenegro, perceive blockchain technology and its implementation in maritime. Through the methodological framework given in [26, 27] we conducted the survey, which included thirteen closed-ended questions, or statements on blockchain adaption. Concerned statements included the following blockchain dimensions: knowledge, infrastructure, standards, experts, diverse stakeholders, government and regulatory policy, social influence, loss of jobs, computing and storing capacity, complexity, opportunistic behavior, sharing information, and security. The respondents were from maritime companies, agencies, research organizations, governmental bodies, insurance companies, and universities. They are from the executive management level in industry and governmental bodies, and active researchers, professors, and lecturers from universities (10 from South Africa and 10 from Montenegro, all with more than 5 years of research experience). The respondents have had to express their (dis)agreement with the proposed statements via Likert 1–5 scale, where 1 represents the lowest level of (dis)agreement, and 5 the highest level of (dis)agreement. The rest of the offered numerical values are respectively in-between these two extremes. The statements and average values of assessments are given in **Table 7**. If the average score per group of respondents is between 1 and 2.5, then the level of (dis)agreement is "low"; if it is between 2.6 and 3.5, then the level of (dis)agreement is "moderate", and if it is between 3.6 and 5, then it is "high".

Five statements with the highest "agree" and "disagree" assessment rates are categorized in different PESTEL (political, economic, social, technological, environmental, and legal) dimensions, along with their rank (**Table 8**).

Statement	Agree	Disagree
S1: The level of awareness and knowledge of BC affects its adoption.	▲ (High)	▲ (Low)
S2: The BC adoption is affected by the availability of the infrastructure and functionality to integrate and interoperate within and across the business ecosystem.	▲ (High)	▲ (Low)
S3: Standardization and ensuring smooth interoperability is necessary, otherwise, BC can make things difficult instead of making them easier.	▲ (Moderate)	▲ (High)
S4: The BC adoption is affected by the availability of skilled and expert resources.	▲ (High)	▲ (Low)
S5: The BC adoption is affected by a large number of stakeholders, with different mindsets, organizational culture, and working habits.	▲ (Moderate)	▲ (Moderate)
S6: The BC adoption is increased by favorable government and regulatory policies.	▲ (High)	▲ (Low)
S7: Social influence positively affects the behavioral intention of using BC.	▲ (Low)	▲ (High)
S8: A perception that BC implementation might lead to loss of jobs can be an obstacle in its adoption.	▲ (High)	▲ (Moderate)
S9: Development in storage, computing, and cloud infrastructure will affect BC adoption.	▲ (High)	▲ (Moderate)
S10: The BC adoption reduces opportunistic behavior (opportunistic behavior means maximization of economic self-interest and occasioned loss of the other partners).	▲ (Low)	▲ (High)
S11: The BC adoption is reduced if the information is not shared by the partners, while some stakeholders are hesitant to share information considering it is a competitive advantage.	▲ (High)	▲ (Moderate)
S12: Privacy and security of models and data need to be ensured, as BC technology is still immature and vulnerable.	▲ (Moderate)	▲ (Moderate)
S13: Blockchain offers a high level of complexity and observability at the same time.	▲ (Low)	▲ (High)

Legend: ▲ High (dis)agreement; ▲ Moderate (dis)agreement; ▲ Low (dis)agreement; BC – blockchain
Source: Own.

Table 7.
The assessments of blockchain adoption.

The respondents consider awareness and knowledge about blockchain as a social dimension of the highest importance for its adoption in maritime. This is understandable, since knowledge is the biggest asset; the only one which grows with exploitation during the time. On the second place is infrastructure, which falls under technological dimension. This is reasonable since, without it, blockchain adoption is practically impossible. In the third places are favorable government and regulatory policies that fall under political and legal dimensions. This is of crucial importance since economic development in South Africa and in Montenegro is controlled by the government. On the fourth place is experts' knowledge, which belongs to the social dimension of PESTEL model, and which is to a certain extent connected with awareness and knowledge, but it can be outsourced in the case of its lack, and under the assumption that awareness and general knowledge about blockchain are in place. The fifth place is the hesitancy of sharing information among the parties, and it falls under both economic and environmental dimensions of PESEL. This is understandable, since once blockchain becomes well established; the impact of this issue will be reduced.

The highest disagreement is observed regarding 'simultaneous' presence of blockchain complexity and observability. Majority of the respondents show suspicion

P Political	E Economic	S Social	T Technological	E Environmental	L Legal
Respondents "Agree"					
*Favorable government policies (rank 3)	*Hesitancy of sharing information (rank 5)	Awareness and knowledge about BC (rank 1) Skilled and expert resources (rank 4)	Infrastructure (rank 2)	*Hesitancy of sharing information (rank 5)	*Favorable regulatory policies (rank 3)
Respondents "Disagree"					
	Reduction of opportunistic behavior (rank 2)	Social Influence (rank 3)	Complexity and observability (rank 1) *Standardization (rank 4) **Ensuring privacy and security (rank 5)	**Ensuring privacy and security (rank 5)	*Standardization (rank 4) **Ensuring privacy and security (rank 5)

- Constructs marked with "" correspond with two different PESTEL dimensions.*
*- Constructs marked with "**"correspond to three different PESTEL dimensions.*
Source: Own.

Table 8.
PESTEL analysis.

regarding this paradox. Furthermore, the respondents do not agree with the statement that blockchain will reduce opportunistic behavior. South Africa and Montenegro are countries that are for decades in transition and suffer from the permanent reproduction of crisis. Consequently, the responders' rather skeptic attitude towards this statement is completely understandable. The social influence is in the third place. The respondents do not believe that society can affect considerably the implementation of this advanced technology, and this belief is based on their experiences from transitional settings. The statement, which deals with the standardization issue, is "negatively" assessed, but it might be the case due to the experts' belief that standardization must be achieved and that it cannot as such threaten blockchain key advantages. The need for ensuring privacy and security is assessed negatively. This means that some respondents strongly disagree with the statement that blockchain technology is still immature and vulnerable. Due to their response, one can conclude they believe that blockchain technology is at a high level of development and that is less vulnerable than it can appear due to its complexity and deployment at a global scale. This construct can correspond with technological, environmental, and legal PESTEL dimensions at the same time.

Since we collected only twenty survey responses, further research should include in-depth interviews or a survey upon a larger poll of experts and profound discussion on the respondents' assessments, including comments and suggestions. In addition, the following investigation in the field should include experts from other developing and transitional countries (besides South Africa and Montenegro), including a longitudinal approach. Building new knowledge and transfer of existing one on blockchain technological and other crucial dimensions are necessary, particularly in developing countries, which suffer the lack of skilled personnel and expert knowledge, dominantly in the technological domain.

4. Conclusion

Maritime is lagging behind other areas of business and industry in terms of digitization. This is especially true in developing countries. Therefore, in this chapter, the results of two case studies on digitalization in maritime, conducted in developing countries, are presented, along with some comparative analyzes with developed countries in certain areas. These studies were conducted on a relatively small sample and in the future, longitudinal studies should be conducted on a larger sample and in a larger number of countries in transition. The results should be presented to ICT&S designers in developed countries in order to find solutions for intelligent design of ICT&S in maritime in perspective. These systems should provide benefits for all. If we strive for sustainable development, then it must ensure development for all. Entire continents cannot be excluded from the plans for further development of ICT&S, artificial intelligence, and virtual reality. This, of course, requires a broader platform, which includes standardization of ports and ships as basic maritime structures onto which ICT&S are built.

General knowledge about new technologies in developing countries exists, but there is a lack of expertise when it comes to hardware, software, network architecture, cyber safety, security, etc. Furthermore, developing countries suffer lack of funds for investments in advanced ICT&S. These countries are excluded from the strategic plans on the development and implementation of new ICT&S solutions. The governments should consequently invest more into attaining sound knowledge in contemporary ICT&S, including disruptive technologies like blockchain and its implementation in global supply chains. Standardization in maritime is the key enabler of faster modernization. In addition, profound research of internet connectivity at sea is of crucial importance for further development of IoT applications in maritime as the key condition for achieving equal presence of ICT&S applications on land and at sea. Maritime industry and high(er) maritime education institutions should cooperate much more closely to alleviate the gap between developing and developed countries in terms of achieving a higher level of graduates' employability, while the sphere of digitalization can provide many opportunities in this respect. Rising awareness about digital transition in maritime and opening discussions among professionals and academics should become common practice. The involvement of legislatures and tight collaboration of key stakeholders in maritime emerging economies are necessary regarding harmonization of ICT&S deployment across global maritime ecosystem and overcoming the existing gap in digitalization.

Author details

Sanja Bauk
Durban University of Technology, Durban, South Africa

*Address all correspondence to: bsanjaster@gmail.com

References

[1] Tsai HC, Chen K, Liu YY, Shuler JM. Demonstration (DEMO) of radiofrequency identification (RFID) system for tracking and monitoring of nuclear materials. Packaging, Transport, Storage & Security of Radioactive Material. 2010;**21**(2):91-102. DOI: 10.1179/174650910X12628787557739

[2] Tsai HC, Chen K, Liu YY, Norair JP, Bellamy S, Shuler J. Applying RFID technology in nuclear materials management. Packaging, Transport, Storage & Security of Radioactive Material. 2008;**19**(1):41-46. DOI: 10.1179/174651008X279000

[3] Bauk S, Dzankic R. Tracking radioactive materials in sea transportation via RFID technology. In: Proceedings of the 8th Maritime Transport Conference (MT2020); 17–18 September 2020. Barcelona, Spain: UNCTAD; 2020. pp. 91-102

[4] Kapidani N, Bauk S, Davidson IE. Digitalization in developing maritime business environments towards ensuring sustainability. Sustainability. 2020; **12**(21):9235. DOI: 10.3390/su12219235

[5] Bauk S, Kapidani N, Schmeink A, Holtham C. Concerning intelligent ICT exploitation in some maritime business organizations: A pilot study. International Journal of Maritime Science & Technology Our Sea. 2017;**64**(2):63-68. DOI: 10.17818/NM/2017/2.5

[6] Bauk S, Kapidani N, Schmeink A. On intelligent use of ICT in some maritime business organizations. Montenegrin Journal of Economics. 2017;**13**(2): 163-173. DOI: 10.14254/1800-5845%2F2017.13-2.11

[7] Bauk S, Calvo LJA, Schmeink A, Azam S, Mathar R. On deploying vehicular communication at the developing seaport and related innovation success impediments. Transport. 2019;**34**(1):126-134. DOI: 10.3846/transport.2019.7987

[8] Holtham C, Stace D, Courtney N. Balancing technology, business processes and skills: key determinants of firm-level information systems success? In: ITU and LBS Digital Transformations in the Information Society Conference; 1-2 June 2006. Geneva, Switzerland: WHO; 2006

[9] Holtham C, Stace D, Courtney N. Mapping opportunity space: Options for a sustainable E-strategy. Strategic Change. 2004;**13**(5):237-251. DOI: 10.1002/jsc.691

[10] Holtham C, Courtney N. Developing managerial learning styles in the context of the strategic application of information and communication technologies. International Journal of Training and Development. 2001; **5**(1):23-33. DOI: 10.1111/1468-2419.00119

[11] Cambridge Dictionary. Make Your Words Meaningful. n.d. Available from: https://dictionary.cambridge.org/dictionary/english/knowledge [Accessed: 09 Febuary, 2022]

[12] Rockart J F. The Changing Role of the Information Systems Executive: A Critical Success Factors Perspective. 1982. Available from: https://dspace.mit.edu/bitstream/handle/1721.1/2010/SWP-1297-08770929-CISR-085.pdf?sequence=1 [Accessed: 13 Febuary, 2022]

[13] Earl MJ. Information Management: The Organizational Dimension. New York, NY, USA: Oxford University Press; 1998. p. 514

[14] Earl MJ. Management Strategies for Information Technology. London, UK: Prentice Hall; 1989. p. 232

[15] Weil P, Ross JW. IT Savvy—What Top Executives Must Know to Go from Pain to Gain. Boston, MA, USA: Harvard Business Press; 2009. p. 182

[16] Kadiri SE, Grabot B, Thoben KD, Hribernik K, Emmanouilidis C, Cieminski G. Current trends on ICT technologies for enterprise information systems. Computers in Industry. 2016; **79**:14-33. DOI: 10.1016/J. COMPIND.2015.06.008

[17] Sanchez-Gonzalez PL, Diaz-Gutierrez D, Leo TJ, Nunez-Rivas L. Towards digitalization of maritime transport. Sensors. 2019;**19**(4):926. DOI: 10.3390/s19040926

[18] Liu J, Zhang H, Zhen L. Blockchain technology in maritime supply chains: Applications, architecture and challenges. International Journal of Production Research. 2021:1-15. DOI: 10.1080/00207543.2021.1930239

[19] World Economic Forum. Enabling Trade, Valuing Growth, Opportunities. 2013. Available from: https://www3.weforum.org/docs/WEF_SCT_EnablingTrade_Report_2013.pdf [Accessed: 09 Febuary, 2022]

[20] Gausdal AH, Czachorowski KV, Solesvik MV. Applying blockchain technology: Evidence from Norwegian companies. Sustainability. 2018;**10**:1985. DOI: 10.3390/su10061985

[21] Ytterstrom J, Lengerg L. What Role Will Blockchain Play Within the Maritime Shipping Industry in Five Years? Gothenberg, Sweden: University of Gothenburg School of Business, Economics and Law; 2019 Available from: https://pdfs.semanticscholar.org/b6d7/cf06c4853ef1745ab4dd12f65f3cc7f5bcb2.pdf [Accessed: 09 Febuary, 2022]

[22] Knirsch F, Unterweger A, Engel D. Privacy-preserving blockchain-based electric vehicle charging with dynamic tariff decisions. Computer Science – Research and Development. 2018;**33**: 71-79. DOI: 10.1007/s00450-017-0348-5

[23] Green E H, Carr E W, Winebrake J J, Corbett J J. Blockchain Technology and Maritime Shipping: An Exploration of Use Cases in the U.S. Maritime Transportation Sector. 2020. Available from: https://www.maritime.dot.gov/sites/marad.dot.gov/files/2020-07/Blockchain%20Use%20Cases%20Final%20Report%20%2820200622%29.pdf [Accessed: 09 Febuary, 2022]

[24] Bhattacharya P, Saraswat D, Dave A, Acharya M, Tanwar S, Sharma G. Davidson I E, coalition of 6G and blockchain in AR/VR space: Challenges and future directions. IEEE Access. 2021; **9**:168455-168484. DOI: 10.1109/ACCESS.2021.3136860

[25] Vardai Z. Top 5 Blockchain Trends for 2022 and Beyond. 2021. Available from: https://forkast.news/top-5-blockchain-trends-with-lasting-impact/ [Accessed: 11 Febuary, 2022]

[26] Kapidani N, Bauk S, Davidson IEA. Developing countries concerns regarding blockchain adoption in maritime. Journal of Marine Science and Engineering. 2021;**9**(12):1326. DOI: 10.3390/jmse9121326

[27] Kapidani N, Bauk S, Davidson IEA. Deploying blockchain in maritime: Developing environments concerns. High Technology Letters. 2021;**27**(9): 942-951

[28] Drljevic N, Aranda D, Stantchev A. Perspectives on risks and standards that

affect the requirements engineering of blockchain technology. Computer Standards & Interfaces. 2020;**69**: 103409. DOI: 10.1016/j.csi.2019.103409

[29] Zhou Y, Soh YS, Loh HS, Yuen KF. The key challenges and critical success factors of blockchain implementation: Policy implications for Singapores maritime industry. Marine Policy. 2020; **122**:104265. DOI: 10.1016/j.marpol.2020.104265

[30] Burgess M. Artificial Intelligence – How Machine Learning Will Shape the Next Decade. London, UK: Penguin Random House; 2021. p. 165

[31] TradeLens. Shipping in the Age of Blockchain. 2019. Available from: https://www.youtube.com/watch?v=Xwqo_fwPEJo&t=1925s [Accessed: 11 Febuary, 2022]

[32] IBM. What are Smart Contracts on Blockchain? n.d. Available from: https://www.ibm.com/za-en/topics/smart-contracts [Accessed: 01 January, 2022]

[33] Jensen T, Henningsson S, Hedman J. Delivering business value with blockchain technology: The long journey of TradeLens. MIS Quarterly Executive. 2019;**18**(4):221-243. DOI: 10.17705/2msqe.00018

[34] IBM and Maersk. TradeLens Documentation: Actionable Doc Flows. 2020. Available from: https://docs.tradelens.com/how_to/actionable_doc_flows/ [Accessed: 10 Febuary, 2022]

[35] NewsBTC – Bitcoin & Cryptocurrency News. First view of Blockshipping's Global Shared Container Platform. 2018. Available from: https://www.newsbtc.com/sponsored/first-view-of-blockshippings-global-shared-container-platform/ [Accessed: 11 Febuary, 2022]

[36] Blockshipping GSCP. Just in Time Containers. 2018. Available from: https://www.youtube.com/watch?v=BUas6vfWLeo&t=85s [Accessed: 11 Febuary, 2022]

[37] Blockshipping GSCP. The GSCP Explained. n.d.. Available from: https://medium.com/@Blockshipping/the-gscp-explained-17804b0b09fc [Accessed: 01 Febuary, 2022]

[38] Blockshipping GSCP. The Four Critical Components of the GSCP 2018. Available from: https://www.youtube.com/watch?v=YIz9-e70hmw&t=55s [Accessed: 11 Febuary, 2022]

[39] Lennart M. Introducing Blockshipping: The Global Shared Container Platform. 2018. Available from https://medium.com/@lennart_7988/introducing-blockshipping-the-global-shared-container-platform-4da87abf0dba [Accessed: 11 Febuary, 2022]

[40] Blockshipping GSCP. The GSCP Flow. 2018. Available from: https://www.youtube.com/watch?v=kMV0A-L-dMw&t=89s [Accessed: 11 Febuary, 2022]

[41] Marine Insight. 7 Major Blockchain Technology Developments in Maritime Industry. 2021. Available from: https://www.marineinsight.com/know-more/7-major-blockchain-technology-developments-in-maritime-industry-in-2018/ [Accessed: 09 Febuary, 2022]

Chapter 7

Autonomous Warehousing: Development and Application of a Maturity Model

Herbert Ruile and Lukas Lichtsteiner

Abstract

Warehouse systems are complex elements in the flow of goods and information in value-creating systems. Their efficiency and effectiveness depend to a large extent on the availability and continuity of information flow and their advanced technical and organizational resources. Recent technological developments in robotics, digitalization and Internet of Things open the pathway towards integrated and autonomous value chain operations. This book chapter describes the development of a measurement model to assess the maturity of autonomous warehouse systems beyond partially automated processes. The model considers technology readiness level, business process maturity as well as organizational capabilities. The maturity model is applied and discussed in a case study.

Keywords: warehouse, internet of things, autonomous systems, maturity model, case study

1. Introduction

Warehouse systems are elementary components of value creation systems to balance demand and supply with an optimal stock of goods. There is no management of value networks in which decisions do not have to be made about the right location, equipment and management of the warehouse. When design and warehouse management is properly understood, it has been shown to result in higher productivity, lower inventory and higher customer and employee satisfaction [1].

The corona pandemic has caused a rapid acceleration of the e-commerce market while highlighting the vulnerability of supply chains and the availability of materials, transportation, and production capacity. With the increasing importance of e-commerce, the demands on the warehouse are rising in terms of product variety, availability, and delivery times for the customer. It is not surprising that investments are currently being made in the management of value creation systems and their central element, the warehouse system [2]. Storage systems take on elementary tasks in the management of material flow: Goods receipt, put-away, storage, picking as well as dispatch. In addition to the actual storage and inventory management, further and diverse services are added, often resulting from a reconfiguration of the supply chain:

including packaging and labelling, quality controls, repairs, assembly, repackaging, reassortment, and more [3, 4].

In addition to high service performance, high warehouse-productivity is always understood as the goal. The most important factors affecting the productivity of a warehouse system include the number of employees and the degree of automation, which in turn interact with each other [5].

The adoption of technological developments is therefore inevitable for logistics productivity. The evolution of information and communication technology in the last decade resulted in the formulation of industry 4.0, Internet of Things or Cyber-Physical Systems, which we here use synonymously for the latest manifestation of digital development. The development of the digital transformation has been called a quantum leap, the 4th industrial revolution, which will radically change our economy, indeed our society [6]. It is therefore understandable that industry 4.0 will also have an impact on SCM, logistics and warehousing [7, 8].

There is a consensus that industry 4.0 pursues the goal of intelligent networking of products and processes in the value chain to increase process efficiency, improve customer service or offer more individualized products and services. We, therefore, follow the industry 4.0 definition applied for supply chain management of [8]:

a. products and services are linked to each other via the Internet

b. the digital linkage allows automated and self-optimized production of products and services, including logistics, without human interaction

c. the value network is controlled peripheral while system elements decide autonomously.

Industry 4.0 is thus expected to make logistics systems more decentralized, self-regulating and efficient. In this context, the core term autonomy is used widely, frequently and consistently. Therefore, we use autonomy also for the warehouse as part of the value chain and logistics. Using the term, it also forces a differentiation from automated warehouses or smart warehouses. Whereas the former relies more on central units with little self-regulation and the latter describes above all the efficiency effect in the warehouse process generated by transparency [7]. In the context of autonomous warehousing, it is thus assumed that the warehouse subsystem can be designed as a decentralized system of the value chain, self-regulating and without human interaction. At least during defined periods of time—a shift extension, additional shift, whole day or weekend—autonomous operation would impact positively the overall equipment efficiency. Identifying and understanding the gap between existing warehouse systems and the ideal situation of an industry 4.0 solution would indicate necessary progress and actions to be taken.

Hence, our research questions are:

a. How to describe the maturity level for warehouse systems?

b. How to assess the individual maturity level?

c. Can action and development paths be derived from this assessment?

We structured the paper accordingly to the recommended methodological approach of maturity model development [9], starting with a literature review, followed by the model-building approach and the model description. The paper continues with the model application on a single case study and closes with preliminary discussion of the results and conclusions for research and management.

2. Literature review

Based on the goal to measure areas of autonomy in warehouses, a literature review was performed. The following chapter will describe the methodological steps that were deployed to collect relevant information.

To begin with, relevant keywords for the search of the literature were derived. As the objective suggests, the main keywords would incorporate 'warehouse' and 'maturity'. Regarding warehousing in general, an important limitation was considered, namely the exclusion of 'data warehouses'. Therefore, the initial search term includes: 'warehouse maturity' – data warehouse. The search term is listed in quotation marks, which allows to search for exact matches in the title of a publication. Next, a search engine of google (scholar.google.com) was selected. This initial search for hits in the title led to only one result. Hence, the authors decided to enlarge the search to other functional areas including logistics (4.0), industry 4.0 and SCM in general. During

Nr.	Authors	Year	Title
1	Warehouse Research and Educational Council	2021	Warehousing & Fulfillment Process Benchmark & Best Practices Guide
2	Salhieh and Alswaer	2021	A proposed maturity model to improve warehouse performance
3	Logistikum Switzerland GmbH	2021	Warehouse Reference Process (Maturity) Model
4	Facchini et al.	2020	A Maturity Model for Logistics 4.0: An Empirical Analysis and a Roadmap for Future Research
5	Zoubek and Michal	2021	A Maturity Model for Evaluating and Increasing the Readiness of the company within the concept of Industry 4.0 with a focus on internal logistics Processes
6	Asdecker and Felch	2018	Development of an Industry 4.0 maturity model for the delivery process in supply chains
7	Competence Centre Medium-sized Businesses	2018	Self-Service to assess the readiness for industry 4.0 in a company
8	Leyh et al.	2016	SIMMI 4.0—A maturity model for classifying the enterprise-wide IT and software landscape focusing on Industry 4.0
9	Sony and Naik	2018	Key ingredients for evaluating Industry 4.0 readiness for organizations: a literature review
10	Santos and Martinho	2018	An Industry 4.0 maturity model proposal
11	Zoubek and Poor	2021	A Maturity Model for Evaluating and Increasing the Readiness of the company within the concept of Industry 4.0 with a focus on internal logistics Processes

Table 1.
Selected studies from performed literature review).

this first intervention, a higher amount of search hits was achieved. In total 11 additional hits were registered for the logistics 4.0 field. Next, 10 hits were counted for the area of industry 4.0 and finally, 5 relevant hits were registered for maturity models in SCM. For these results, titles, abstracts and summaries were analyzed. A second intervention was carried out and a deeper look at the publications revealed shortcomings, which led to the exclusion of 8 sources from logistics (4.0), 6 sources from industry 4.0 and 5 from SCM. On the other hand, a snowball-approach was deployed to investigate more into references made by relevant authors. This route led to 3 more interesting studies (2 from warehousing and 1 from industry 4.0). The final selection encompassed 11 papers that were further analyzed (**Table 1**).

2.1 Maturity dimensions

Since we were interested to find relevant dimensions in which the maturity of warehouses towards autonomy could be measured, we extended a request into our professional network of warehousing and SCM experts. The goal was to develop a reference process for warehousing that allowed to search for commonalities and deviations in the registered search results. These findings would eventually help us to confirm or to discard certain elements from the maturity model. The expert talks helped to confirm initial assumptions and extended the knowledge in this domain. As a summarization, the following essential process-steps were identified: deloading (of trucks), receiving, material handling and put-away, storage control (inventory management), picking, packing, loading (of trucks) and shipping. The expert rounds assisted in uncovering another selection of important functional areas that would complement the initial approach. These are the yard management as well as management of information technology, which is critical for the seamless flow of information, goods and finances. The dimensions identified were then mirrored against the papers to find commonalities and differences.

As shown in **Table 2**, the first comparison of the reference process with existing literature highlights important findings. In comparison to the Maturity Model of WERC [10], it is found that all aspects of the reference process, apart from de-loading and yard management, could be confirmed. Subsequently, the reference processes were investigated regarding the study of Salhieh and Alswaer. **Table 2** demonstrates that less commonalities could be identified. At least receiving, put-away, picking and shipping could be confirmed as overlapping process steps. Unlike the previous model, Salhieh and Alswaer clearly refer to maturity levels and recommend the usage of 5 different maturity levels. Lastly, the reference process was mirrored against the previous work of the authors. Related to an external project, the authors outlined a basic model to describe interactions and dependencies in modern warehousing. The comparison reveals that most of the process steps could be confirmed, apart from receiving, storage control, packing and shipping. As previously, 5 maturity levels were listed.

As outlined in the previous part of this chapter, the analysis was later extended into the area of logistics (4.0). This comparison is based on three studies. The first one covers the development of a maturity model for logistics 4.0 and includes a roadmap for further research [11]. The second paper goes into details regarding a framework for logistics maturity assessments, respectively with a portion that considers internal logistics [12]. Lastly, Asdecker and Felch present the development of maturity model for the delivery process in supply chains [13]. Unlike the previous comparison, the authors proceed without a detailed comparison against the reference process. The results were not fruitful enough and there were not enough commonalities to justify further discussion.

	Reference process	WERC[a] (2021)	Salhieh & Alswaer (2021)
Functional areas	De-loading	n/a	n/a
	Receiving	Receiving and inspection	Receiving
	Material handling and put away	Material handling and put away	Put away
	Storage control	Storage and inventory control	n/a
	Picking	Picking	Picking
	Packing	Packing	n/a
	Loading	Load consolidation	n/a
	Shipping	Shipping	Shipping
	Yard mgt.	n/a	n/a
	IT mgt.	Warehouse mgt system (WMS)	n/a
	Maturity levels	n/a	5

[a]*Warehouse Education and Research Council*

Table 2.
Analysis of commonalities and differences of warehouse reference process in warehousing literature.

Contrasting, the section of literature related to industry 4.0 brought up interesting findings. Firstly, the reference process had to be adapted, since industry 4.0 covers more areas than warehousing alone. Commonly established models describe the vision of industry 4.0 with a variety of aspects: Business models, digital products & services, processes, production, social aspects, organizational factors, IT and digitalization and logistics. The first selected publication originates at the 'Competence centre for medium-sized business of North Rhine-Westphalia' and was later adapted by another company since public funding halted [14]. Next, the paper of Leyh et al. is referred [15]. The scientists consider assessing the IT and software landscapes of enterprises and propose a model for companies to increase their readiness for industry 4.0. Thirdly, the paper of Sony and Naik was selected, in which a literature review provides the basis to discuss the key ingredients for evaluating industry 4.0 readiness in organizations [16]. Fourthly, a study of Santos et al. is referenced, who discuss and propose an industry 4.0 readiness model [17]. Lastly, an additional study by Zoubek et al. was selected. It addresses a maturity model which could assist in evaluating and increasing the readiness within the concept of industry 4.0 while considering a focus on internal logistics processes (**Table 3**) [18].

Like the foregoing section, the goal of this comparison is to identify corresponding and deviating aspects regarding constituent elements of industry 4.0. Considering the information that can be extracted from the readiness assessment of NRW's competence centre, most of the elements seem to reappear. The same applies to the paper of Leyh et al., although the emphasis on digitalization, integration and cross-sectional technology implementation is stronger. Sony and Naik and Santos and Martinho mostly confirm previous findings but underline the importance of smart, respectively intelligent products, services and processes. From the comparison with the more general study of Zoubek and Poor, the emphasis is put on production and logistics as well as information technology.

elements	NRW	Leyh et al.	Sony and Naik	Santos and Martinho	Zoubek and Poor
Business model, digital products & services	Business models	Digital product development	Smart product, smart services	Smart products, smart services	n/a
Processes	Process mgt.	n/a	n/a	Smart processes	n/a
Production	Planning, control of production	n/a	n/a	Smart factory	Production
Social aspects	Human-machine-interaction	n/a	Employee adaptability towards I4.0	Work-force	n/a
Organization, Strategy	n/a	Vertical, horizontal integration	Organization, strategy, Top mgt. Support	Organizational strategy, structure, culture	n/a
IT, Digitalization	IT Systems	Cross-sectional technology	Level of digitalization of organization, extent of digitalization of supply chain	n/a	IT
Logistics	Logistics, distribution, mgt. of public procurement	n/a	n/a	n/a	Logistics
Maturity levels	5	5	n/a	6	6

Table 3.
Analysis of design elements in industry 4.0 literature.

2.2 Maturity levels

Apart from the maturity dimensions that were discussed, a second approach was used to study the maturity levels of different models. While maturity dimensions point to the area where the measurement for maturity will take place, levels are used to assess a certain readiness in a particular area. To perform this assessment, it is important to consider that maturity levels should be chosen coherently and either in a quantitative or qualitative way. This definition also affects the interpretability of the results.

Regarding the first group of papers that were analyzed, mostly similar layouts of maturity levels were found. Due to a paywall, the maturity levels of WERC's maturity model could not be accessed. The research of Salhieh and Alswaer refers to 4 different maturity levels. To assess the maturity of areas like integrated warehouse performance measures, the two scientists propose levels starting from negligible, low, moderate or high. Each of the levels has more details to it, for example, a negligible maturity level would correlate with sub 25% usage or deployment of a certain measurement, while a high maturity level would correspond to the usage of performance measures that is in the range of 75–100%. Next, the investigations of Logistikum Schweiz GmbH resulted in 5 maturity levels. In the Warehouse Reference Process Model, various dimensions are addressed. As an example, the assessment in Yard Management Maturity refers to manual, mechanized, automated, digitally augmented or lastly, intelligent dark. To exemplify, the final intelligent dark maturity level would describe that all the work and services are done in an autonomous way, including autonomously operating robots. Regarding the findings in the industry 4.0 section, mostly consistent levels were found. The competence centre for medium-sized businesses suggests 5 different levels. The starting point is marked by paper transfer of data, transfer of paper data in digital form, general usage of ERP systems, and digital data completeness until the automatic transfer of data. In a similar fashion, Leyh et al. suggest the application of 5 maturity levels. To assess the maturity of the IT landscape the following levels are used: basic digitization, cross-department digitization, horizontal and vertical digitization, full digitization and optimized full digitization. While Sony and Naik would not address the details of maturity models, suitable considerations can be found for the studies of Santos and Martinho as well as for Zoubek and Poor. Both suggest 6 maturity levels in their models. Structurally, they only differ slightly from the previously investigated models. This differentiation is based on the first, initial maturity level, which is congruent for both, as their first levels start at zero actions, respectively zero shares of implemented initiatives.

To summarize this comparative representation, all reviewed maturity models relate to well-established components of industry 4.0. If this comparison is extended to warehousing, some notable differences come to attention. While it is obvious that industry 4.0 maturity models address basic functions like process management, production and logistics, additional elements like social, organizational and technological viewpoints are addressed as well. These elements are rarely represented in functional maturity models, like maturity models of logistics processes or applied technologies in logistics.

3. Model building approach

Regarding the model development, a suitable methodological approach has to be chosen. Comparable investigations refer to the work of De Bruin et al. [9]. In their seminal paper, the scientists presented an often-cited approach that assists in developing specific maturity assessment models.

The approach is based on six subsequent phases: (1) scope, (2) design, (3) populate, (4) validate, (5) test and deploy and (6) maintain. Since this study is not a longitudinal study, only the first five phases will be used.

Phase 1: Scoping in the first phase of scoping, a decision must be made whether the model will address general or domain-specific use, which determines the scope and boundaries of the suggested model [9]. Apart from this decision, it is important to consider and include further stakeholders in the development of the model. This should ensure that possible benefits that result from the development or result from the use of the model can be shared with experts and vice-versa, experts can help and contribute to the model and its development stages. The exchange with science and industry is of great importance because it allows to build on existing knowledge and insights from previous research.

Phase 2: The design phase centres around five subsequent criteria, that determine the further layout of the model. They are intended to clarify the audience, the method of application, the driver of application, respondents and the application itself. For this study, the audience is mainly warehouse managers because they are directly involved in initiatives regarding the organizational and technological development of their facilities. Furthermore, the second audience of interest are consultants and auditors, who are often involved in guiding and accompanying warehouses. The principal method of application will be mostly based on structured interviews. Based on the clarification of the reason, why such a model should be applied and for whom it will be developed, the next chapter addresses the remaining design aspects.

Phase 3: Populate: The next phase is centred around the population of the maturity models in terms of content and requires the description of model components and model subcomponents [9]. This description clarifies what content needs to be measured for any given component or subcomponent. According to DeBruin et al., various approaches are suited to define the contents of each subcomponent. For example, a thorough literature review could be suited, as well as empirical approaches such as stakeholder interviews, surveys, focus groups and in-depth case studies. For the present study, a combination of approaches was chosen. Firstly, a literature review was conducted to identify basic components and subcomponents. Secondly, the study used individual expert talks and semi-structured interviews to validate the findings and to check if certain aspects need to be further adjusted. The results of the literature review are discussed in a previous chapter. The feedbacks resulting from the expert talks largely confirmed initial viewpoints and assisted in validating the principal assumptions of the maturity model.

Phase 4: Model validating: The validity and reliability of the maturity model in scope are in the focus of the fourth phase of model development, according to DeBruin et al. validity and reliability are important building blocks to ensure and strengthen the relevance and rigor of the model [9]. While the validity of the model is supposed to secure the correlation between factual and intended measurements, the reliability addresses if the obtainable results are accurate and repeatable. As in previous sections, referring to DeBruin et al. reveal different approaches to ensure such requirements. Surveys, interviews or literature reviews are among the options to be used in this regard. For the present study, all maturity dimensions and maturity levels are grounded in previous research publications and were validated in expert interviews. Therefore, the validity and reliability are confirmed.

Phase 5: Test and deployment. Within the last consecutive phase of model development guidelines [9], the deployment of the model is addressed. Following the guidelines of DeBruin et al., this phase aims to clarify the generalizability of the model pursued. This can be achieved by applying the model within suitable case studies. DeBruin et al. refer to two separate approaches. In the first place, it is suggested to test the model within an audience consisting of stakeholders, who were directly or indirectly involved in the development of the model itself. Secondly, the generalizability can be extended by discussing the model with an audience that is not part of a stakeholder group, respectively is external to the domain of warehouse maturity research. By pursuing these two steps, the general acceptance of the model can be reflected and confirmed. For the research at hand, the testing of the model was done by discussing it with members of a specific focus group that is committed to develop the warehouse of the future and consists of various warehouse operators, consultants and solutions providers with extensive experience in this domain. As a result of discussions, it could be found that the majority agrees with the proposals made by the authors. These discussions not only helped to ensure acceptance and demand for the model but also assisted in enriching the initial building. The focus group mentioned above also involves research partners from two universities of applied sciences. From each of the partners, one person who was not involved in the model development was asked to review the model and provide feedback to the authors. Like the first group, the authors could not learn about contributions that would question the current state of the model. To summarize, this initiative led to further confirmation of the model proposed.

4. Definition of the maturity model

As discussed in the previous chapters, the search for commonalities and differences in antecedent maturity assessment models mainly led to findings that underline the process-oriented view. By considering this view, the initial model will assess the most important areas of action within a warehouse, such as unloading, receiving, put-away, storage, picking, packing, loading and shipping. During the investigations into maturity dimensions from other domains such as logistics and industry 4.0, more contributing aspects are identified. Notably, the maturity models from the domain of industry 4.0 extend the initial process model. Regarding the first group of findings, the results that were obtained by comparing warehouse and industry 4.0 models pointed to additional important features, such as a general process-oriented approach, people, technology adoption and implementation and lastly organizational aspects. These features are considered relevant because they refer to the approach that industry 4.0 concepts take. These concepts are important for the present research, as this study is interested in assessing warehouse maturity regarding the realization of a fully autonomously operating warehouse – a concept deeply grounded in the environment of industry 4.0.

By combining these findings, it becomes apparent that the classical approach of a two-dimensional view, that most maturity models incorporate, starts to evolve. Currently, each dimension added to the model requires more attributes to be considered. The perspective of process control by management, the relevance of people in focused processes and the organization itself have consequences for each of the maturity dimensions. It can be safely assumed that for each of these elements, individual maturities can be assessed and therefore addressed for improvements. Hence, the proposed maturity model includes for the first dimensions process-related 'dimensions' and for the second dimension, intersectional factors like process and people management, technology applied and the organizational design (**Figure 1**).

As presented in the previous section, mostly consistent findings pointed to the usage of either 5 or 6 maturity levels. As explained by DeBruin et al. [9], it is important that all the levels are clearly defined, distinct and logically progressing from one to another. Furthermore, the authors underline the importance of all requirements and

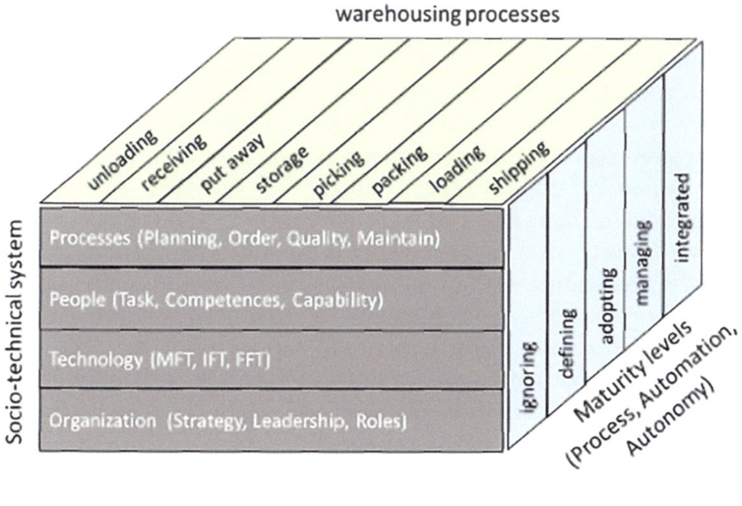

Figure 1.
Warehouse maturity assessment model.

Maturity level	Description
Ignoring	No awareness regarding needs for integration of advanced technology into operations. Missing knowledge about potentially beneficial advanced solutions, that could enhance flow of goods, information, finances.
Defining	Need for integration is acknowledged, but missing knowledge in terms of implementation. Knowledge of advanced solutions available, but missing knowledge in terms of application.
Adopting	Basic steps for integration of advanced technology initiated. Implementation of some advanced solutions, that enhance the flow of goods, information, and finances.
Managing	Integration is driven forwards and affects most business areas. Most of the solutions currently available are implemented to enhance and improve operations.
Integrated	Top level of integration is established, synergies show their full potential. All advanced solutions on the market are implemented, resulting in a seamless and optimized flow of material, goods, and information.

Table 4.
Description of generic maturity levels.

how the measurements are detailed. In this reasoning, maturity levels are supposed to represent a certain degree of maturity in their respective dimension and allow for improvements in a chosen field of interest. To operationalize the measurements, every level needs respective denotation and furthermore, a general description. The present approach proposes 5 generic and logically succeeding maturity levels, which are portrayed in the following table (**Table 4**).

Reflecting on the foregoing sections, the above illustration shows the warehouse maturity assessment model. In the upper-top area, the warehouse process-oriented dimensions are recognizable, while on the right-hand side the maturity levels are shown vice-versa. In comparison to other existing maturity models, the differences are becoming clearer as illustrated by the bottom area, which additionally integrates the socio-technical viewpoint. A socio-technical system usually considers three main building blocks: a technological, an organizational and a workforce-related, respectively human-oriented one. To complement the process-oriented maturity dimensions and interlink them with the socio-technical system, this study considers an additional, generic process management layer for this model. This link between the shopfloor-related warehouse processes and the management of those processes allows for a more complete and holistic analysis of other important aspects of warehouse maturity. Any process in each warehouse setting builds on organizational elements, technological equipment and foremost on people. Furthermore, processes can be characterized by flows of different types. The flow of goods (MFT) often marks a starting point, followed by the flow of information (IFT) and the flow of finances (FFT). Another important aspect to consider is the way organizations document their own processes. Such documentations mark an essential orientation for employees and managers who are involved in developing and improving current processes.

Since this study aims to measure maturity in various areas of warehouse operations, the above-listed maturity dimensions need to be adjusted individually. The required adjustments will still follow the above-defined generic maturity dimensions but will slightly differ to capture the specific nature of maturity in selected warehouse maturity dimensions. As per follows, the required adjustments in the case of one exemplary maturity dimension, the process of unloading trucks, shall be presented.

As **Table 5** shows, the individual adjustments span over all sub-elements that were identified to be important per singular process-step. **Table 5** extends these denotations by addressing the sub-element of organization, where three more important elements were found.

As listed in **Table 6**, the extended denotations for the maturity levels of the sub-element of organizational aspects are detailed.

Maturity dimension	Stage 1	Stage 2	Stage 3	Stage 4	Stage 5
Process documentation	Not documented	Partial analogue documentation, no centralized storage	Centralized storage, analogue documentation	Partially digitized, slight deviations in analogue vs. digital storages possible	Fully digitized, centralized, always available via dedicated platform solution (e.g., QMS[a])
Material flow	Manual unloading, manual transfer	Mechanically supported unloading (e.g., with forklifts)	Manual unloading to conveyor systems	Partially automated unloading (i.e., with automated forklifts and conveyor systems)	Automated, resp. autonomous unloading with robotic equipment
Information flow	Manual processing of information, data islands	Data available in ERPs, no automated exchange of data between functional areas	Flow of data widely automated, available for pre- & anteceding process steps	Flow of data end-to-end available, automatic exchange of data between stakeholders	Digitized, fully automated exchange of data, ongoing use for optimizations
Technology	No automation, service provision 100% manually	Partial automation, human activities supported by mechanical equipment	Limited automation, human-machine-cooperation	High automation, humans in charge of controlling only	Autonomous service provision, without any human interventions
People	Semiskilled, no technical formation, no technical further education	Completed traineeship, and/or professional maturity, no technical further education	Completed traineeship and/or professional maturity, technical further education done in last 3–5 years	Completed traineeship and/or professional maturity, recently completed technical further education	Completed traineeship and/or professional maturity, more than one recent technical further education, additional recent specific diplomas

[a]*Quality management system*

Table 5.
Exemplary description of maturity dimensions for the process "unloading of trucks".

Organization	Stage 1	Stage 2	Stage 3	Stage 4	Stage 5
Coordination	Little coordination between roles and functions	Ad hoc coordination, if initiated and/ or required	Sporadic meeting for coordination	Matters of coordination regularly discussed	High coordination of roles and functions
Culture	Absent company culture, potentials for improvement are not seen	Barely opened culture, potentials for improvement seen, but not discussed	Rather open culture, potentials for improvements seen and sporadically discussed	Mostly open culture, potentials for improvement are recognized and discussed regularly	Very open culture, systematic reviewing of potentials for improvement, implementations to follow
Resources	Scarcity of resources, pressing needs cannot be satisfied	Scarcity of resources, required resources are granted rarely	Resources obtainable but complicated procedures, usually delayed	Resources obtainable through pre-defined procedures, rather timely	Resources can be easily obtained, pre-defined procedures available, very timely and direct

Table 6.
Description of maturity dimensions for the process 'unloading of trucks'.

5. Case study

Sample firm: As an initial partner for the first application of this maturity assessment, a company was selected which is engaged in providing warehouse facilities for pharmaceutical products. This company's business is best described as a 'pre-wholesaler'. According to this business type, the selected company is providing logistical services for pharmaceutical companies. Typically, this includes warehousing and transportation of pharmaceutical products from the manufacturers to wholesalers, hospitals and in some cases to pharmacies directly. The business of pre-wholesaling is special, as the supplier of such services does not take title to the products they are storing, and ownership stays with the manufacturer until delivery is complete. Furthermore, pre-wholesalers do not maintain customer relationships with the intended recipient of the product, but with the manufacturers who pay a fee for the services used.

This company describes itself as very engaged in the digital transformation of its operations and has already undergone some small steps and test projects. Therefore, this company seems to be well suited for the application of the maturity model proposed. The company is described by roughly 100 million dispatches p.a. and about 200 employees. Since its establishment, the company is engaged in improving its customers' storage needs and providing various value-added services.

Data collection: To conduct an initial maturity assessment, the authors had to develop a questionnaire. As explained in the foregoing sections, the main building blocks of this questionnaire include warehousing processes as the main rationale for describing the maturity dimensions and corresponding maturity levels. In total, the questionnaire considers six questions per dimension. Regarding all dimensions (7), the total sum of questions equals 42 questions. The questions are directed to learn more about the specific state of maturity in the dimensions along the warehouse processes and can be detailed as follows:

- What is the maturity level of the documentation in this process? Is it possible for employees to access the documentation at any time if questions need to be resolved or the process is going to be reflected for improvements?

- What stage does the maturity have regarding the flow of material in the process of unloading trucks? Is the manual handling of material overweighing or do you already use assisting technologies to ease the effort and streamline the process?

- How would you describe the maturity of the flow of information? Does the organization tend to manual workings and share handwritings between functions or is it already more automated and using electronic interfaces?

- Which technologies are used in this process? We would like to identify all the technologies in place and assess individual maturity levels for each of them.

- To be able to assess the maturity of the workforce, please provide the state of formation and technical expertise for each of them?

- How would you describe the maturity of the organization itself in this subprocess? This question includes and addresses three more aspects: the coordination between roles and function, the state of the company culture and the availability of resources.

The qualitative results that would be collected by this survey allowed the achievement of the principal objective.

Findings: The semi-structured interview allowed for capturing information not only tied directly to the questions involved but also information that helps to understand contextual aspects.

Regarding the first question, respondents stated that the entire process documentation for the unloading process is stored digitally and accessible through a quality management system, which equals the highest level of maturity (5 of 5). Furthermore, the entire documentation is updated once per year through a predefined and participatory process. Every 3 years, all documents and instructions are subject to revision, guided by an external auditor. As assumed, the process documentation is indeed being used for internal advanced training. Apart from regulated measures, spontaneous initiatives like CIPs[1] or unexpected other events could trigger an update within the documentation. The company acknowledges room for improvements regarding the use of digital signatures for increasing safety and accountability.

As for the second item of the questionnaire, the respondents explained that the flow of material within the process of unloading trucks is mainly mechanically supported by equipment such as forklifts. Regarding this information, the company reaches 2 out of 5 possible maturity degrees. The respondents further explained that they were pursuing a small number of initiatives to test a higher degree of automation using automated guided vehicles. In view of the maturity of the flow of information, diverse results were registered. Firstly, the interviewees pointed out that there is still a data island within the company since a considerable amount of data is not digitized at all and is handled by using and sharing handwritten memos (maturity level 1). At the

[1] Continuous improvement process.

same time, the company relies on an enterprise resource planning system (ERP) that enables and supports business processes and communication with clients (maturity level 2). A further nuance could be registered, as some limited operations are also using streamlined electronic data interchange (EDI; maturity level 3). In turn, no unequivocal maturity level could be identified but rather a distribution over three different maturity levels became apparent.

The next item covered the maturity levels of technology applied in the process of unloading trucks. The respondents listed the different technological solutions currently in use, while the researchers connected them separately with maturity levels. A diverse picture began to emerge, starting with equipment of types such as forklifts, video surveillance and entry doors – which were counted in the category of least maturity, as they were all manually operated. The second level of maturity could be seen in equipment such as a packing machine, scanning devices and several displays (used for showing order-related information to shop-floor workers) as well as a semi-automated, web-based IT-system for the provision of slots for incoming trucks. In this category, the technologies assisted humans in their work but were not yet fully automated.

Regarding the next section, the maturity of the workforce was investigated. Since there was a total of 18 employees involved in this subprocess, the assessment turned out to be rather difficult. After several talks, the assessment team and the respondents concluded, that the 4th maturity level (5 as highest level) would be suited the most as it applies to more than 90% of the 18 employees. It was clarified that the job profiles were diverse and differentiating requirements were in place. The 18 employees can be distributed to different sub-areas within the unloading process, 2 of them to the area of ambient storage, 3 workers are in charge for regular daily business (e.g., administrative work), 6 persons to the area of goods receipt and 7 more were engaged with shopfloor activities, such as operating forklifts.

The last section of the semi-structured survey addressed the maturity of the organization, respectively selected important organizational features. The first item of interest points to the importance of internal coordination between functions and roles. The interviewees declared that most of the coordination takes place in the form of rather irregular internal meetings. Although in some cases, the meetings are scheduled and follow pre-defined protocols. In sum, the maturity level for the coordination falls between levels 3 and 4. The second item of this category directs to the state of the company culture. Like the foregoing assessment, the respondents explained that the company culture is mostly open, where potentially interesting external contributions—such as innovative technology—are openly acknowledged, while the discussion on such aspects is rather rare. The respondents were not completely unambiguous, and some pointed out that the assessment might also hit the 4th level of maturity here, which corresponds with slightly increased awareness towards valuable contributions. Finally, the item of resource management and availability was highlighted. Again, it was laid out that in most cases, the request for additional resources is often characterized through rather complicated procedures but could also be sped up, if there is a critical demand to meet.

6. Discussion

The main goal of this research was to develop the principal building blocks of a warehouse maturity model and to apply them to a real case. To reach this objective, we initially performed a literature review to explore available maturity models that address warehouse maturity towards automation and, more importantly, autonomy. We found that there is a relative scarcity of models to build on. Therefore, our model is based on warehouse processes as an orientation for developing maturity dimensions. Because the concept of industry 4.0 postulates a holistic transformation of business models a focus on process maturity does not match with an integrated approach. Aspects such as technology management, the way the organization works internally and the importance of the workforce will have an impact and should be considered when assessing maturity in warehouses. By the integration of the industry 4.0 model as well as organizational maturity, we enriched the process model in order to achieve an integrated maturity model for the autonomous warehouse. To apply and test the maturity model, we used a single case study of a midsized logistics service provider. The main business activity is pre-wholesaling for pharmaceutical products; the main service is warehousing for manufacturers of pharmaceutical products and related value-adding logistics services. For the assessment itself, we used document analysis, semi-guided interviews with operators and management, process walks as well as interactive workshops to capture in-depth information, and to discuss and interpret the findings. The assessment was performed by three researchers in one day workshop. In the maturity assessment conducted, the process was able to identify and confirm both the current state and the development needs for an autonomous warehouse system.

Our observations allowed for a multi-dimensional view of the maturity of the warehouse system of this company. The predefined levels were easy to understand and sufficient in differentiating levels. The application procedure ensures a concise assessment providing further insights into how the company handles challenges and opportunities as well as how to meet customers' increasing requirements.

As the maturity levels were pre-described, the assessment is a semi-qualitative approach. Therefore, it gives room for individual understanding and interpretation. The selection of interview partners, as well as the workshop structure, allows to establish an integrated understanding of the actual situation. As industry 4.0 postulates an integrated transformation, which affects not only technology but also people and organization as well, the model was not able to describe and evaluate the inter-relationship between technology, organization and people maturity. It seems to be obvious that technology impacts organizations and people, but the knowledge and design for it require further investigations.

7. Conclusion

In the last decade, many technological innovations became readily available for warehouse operators across the globe. It is a challenge for any warehouse operator to select the most fitting and efficient technologies to stay competitive among other factors on the market. The change towards networked and intelligent operations is ongoing and the pressure to innovate is one of many driving factors. Maturity models can mark an important waypoint in this challenge since they allow an efficient qualitative as-is analysis of operations and are able to support the development of integrated roadmaps. In this light, the authors aimed to develop and test such a maturity model in a practical environment.

For the development of the maturity model, we built on the recommendations by DeBruin et al. [9] and considered relevant literature and expert interviews to validate and ensure reliability and rigor of the maturity model. By using a single case study, we initially deployed and tested the maturity model and the application procedure for the good receiving process from unloading trucks to storing the goods.

The maturity model for the future of warehousing integrates maturity dimensions of process, technology and organization, which is postulated by industry 4.0. The model describes in detail the maturity in five levels, in two dimensions and subordinated categories. In total 32 elements.

The application of the model in a single case study results in a correct, comprehensive and intuitive reproduction and representation of the actual warehouse situation landscape.

The new model includes and considers aspects from industry 4.0 approaches such as autonomy and organizational design. By this, our model outlines a more holistic approach to the digital transformation of warehousing as part of an autonomous value chain. We have tested the model on a single case study in Switzerland, which does not allow to make assumptions on any other industries, company sizes or countries. Even we may recognize low maturities and gaps for each company, there is no indication of reasoning to invest for achieving a higher level. The context-specific application of the model may rise drivers and hurdles for continuous warehouse developments.

For warehouse management as a socio-technical system that considers people, process, technology and organization, the application of the model gives opportunities for improvement in a holistic way: maturity in logistics processes, technology and organization. Using this model as a starting point to design the transformation roadmap for the warehouse of the future gives awareness about the interrelationship in multiple dimensions. Even if we do not know exactly the interrelationships between the dimensions, it makes clear that digital transformation in logistics is more than implementing technology. It will affect management processes as well as roles, rules and according technical and collaborating skills.

Author details

Herbert Ruile* and Lukas Lichtsteiner
Logistikum Schweiz GmbH, Altdorf, Switzerland

*Address all correspondence to: herbert.ruile@logistikum.ch

References

[1] De Koster RB, Johnson AL, Roy D. Warehouse design and management. International Journal of Production Research. 2017;**55**(21):6327-6330

[2] Research and Markets. "Research and Markets: The Worlds Largest Market Research Store," 16 February 2022. [Online]. Available: https://www.researchandmarkets.com/reports/5397084/global-warehouse-automation-market-insights-and [Zugriff am 16 February 2022]

[3] Higginson J, Bookbinder J. Distribution centres in supply chain operations. In: Logistics Systems: Design and Optimization. Boston MA: Springer; 2005

[4] van Hoeck R. Reconfiguring the supply chain to implement postponed manufacturing. The International Journal of Logistics Management. 1998;**9**(1):95-110

[5] de Koster RB, Balk B. Benchmarking and monitoring international warehouse operations in Europe. Production and Automation Management. 2008;**17**(2):175-183

[6] de Man J, Strandhagen J. An INdustry 4.0 research agenda for sustainable business models. Procedia CIRP. 2017;**63**:721-726

[7] Barreto L, Amaral A, Pereira T. Industry 4.0 implications in logistics: An overview. Procedia Manufacturing. 2017;**13**:1245-1252

[8] Hofmann E, Rüsch M. Industry 4.0 and the current status as well as future prospects on logistics. Computers in Industry. 2017;**89**:23-34

[9] De Bruin T, Freeze R, Kaulkarni U, Rosemann M. Understanding the main phases of developing a maturity assessment model. In: ACIS 2005 Proceedings. Australian Conference on Information Systems (ACIS): 2005

[10] Warehousing Education & Research Council. [Online]. 2021. Available: https://werc.org/page/Benchmarking [Zugriff am 21 Juli 2021]

[11] Facchini F, Olesków-Szłapka J, Ranieri L, Urbinati A. A maturity model for logistics 4.0: An empirical analysis and a roadmap for future research. Sustainability. 2020;**86**:1-18

[12] Zoubek M, Simon M. A framework for a logistics 4.0 maturity model with a specification for internal logistics. MM Science Journal. 2021:4264-4274

[13] Asdecker B, Felch V. Development of an industry 4.0 maturity model for the delivery process in supply chains. Journal of Modelling in Management. 2018;**4**:840-883

[14] Saarbrücken K. Digitaler Reifegrad ihres Unternehmens. 16 February 2022. [Online]. Available: https://kompetenzzentrum-saarbruecken.digital/selbsteinschaetzung/. [Zugriff am 16 February 2022]

[15] Leyh C, Bley K, Schäffer T, Förstenhäusler S. SIMMI 4.0—A maturity model for classifying the enterprise-wide IT and software landscape focusing on industry 4.0. In: 2016 Federated Conference on Computer Science and Information Systems (FedCSIS); 2016

[16] Sony M, Naik S. Key ingredients for evaluating industry 4.0 readiness

for organizations: A literature review. Benchmarking: An International Journal. 2020;**27**(7):2213-2232

[17] Santos RC, Martinho JL. An industry 4.0 maturity model proposal. Journal of Manufacturing Technology Management. 2020;**5**:1023-1043

[18] Zoubek M, Poor P, Simon M. A maturity model for evaluating and increasing the readiness of the company within the concept of industry 4.0 with a focus on internal logistics processes. In: Proceedings of ISER International Conference. Moscow, Russia: Institute for Technology and Research (ITRESEARCH); 2020

Chapter 8

Fluid Inventory Models under Markovian Environment

Yonit Barron

Abstract

Today's products are subject to fast changes due to market conditions, short life cycles, and technological advances. Thus, an important problem in inventory planning is how to effectively manage the inventory control in a dynamic and stochastic environment. The traditional Economic Order Quantity (EOQ) and Economic Production Quantity (EPQ) both are widely and successfully used models of inventory management. However, both models assume constant and fixed parameters over time. Unfortunately, most of these assumptions are unrealistic. In this study, we generalize the EOQ and EPQ models and study production-inventory fluid models operating in a stochastic environment. The inventory level increases or decreases according to a fluid-flow rate modulated by an n-state continuous time Markov chain (CTMC). Our main objective is to minimize the expected discounted total cost which includes ordering, purchasing, production, set up, holding, and shortage costs. Applying regenerative theory, optional sampling theorem (OST) to the multi-dimensional martingale and fluid flow techniques, we develop methods to obtain explicit formulas for these cost functionals. As such, we provide managers with a useful framework and an efficient and easy-to-implement tool to coop with different demand–supply patterns.

Keywords: Inventory/production, Markov chain, Fluid flow, Renewal theory, Martingales, EOQ, EPQ

1. Introduction

An important problem in inventory planning is how to effectively manage the inventory control in a dynamic and stochastic environment. Particularly, today's products are subject to fast changes due to market conditions, short life cycles, and technological advances. The uncertainties in the supply-chain hierarchy make inventory control very challenging and the structure of the optimal policy is still unknown; thus, both researchers and practitioners are focusing on relatively simple control policies. This focus is especially true in practice where there are uncertainties in the production rate, demand rate, and lead time. The specification of the production rates, order quantities, the storage capacity and the backlog possibilities has to take into account the costs for ordering, production, the holding of inventory, backlogging, and lost sales. For example, for a factory Manufacturer, fixing the production rate at some high level or ordering a large quantity avoids backlogs but may cause high production/order cost and inventory costs; however, fixing the production rate at some low level, or ordering many but small quantities may lead to severe backlogging costs and the loss of sale opportunities. An inventory control policy governs inventory replenishment decisions by specifying when and how many items should be ordered or produced, thus, the planning of such a supply chain subject to market uncertainty is challenging in terms of dealing with uncertainties and variations.

The traditional Economic Order Quantity (EOQ) and Economic Production Quantity (EPQ) both are widely and successfully used models of inventory management (see Nahmias [1]). In the classical EOQ model the demand occurs at a constant rate, and each time the inventory level hits level 0 an immediate and fixed order of size Q is placed. In the classic EPQ model, every production cycle is composed of ON and OFF deterministic periods. There exists a predetermined level, say Q, such that the system is ON and the inventory level increases from level 0 up to level Q. When level Q is reached, the production is stopped and the inventory decreases down to 0. The time it takes from Q to 0 is the OFF period. Notice that the EOQ model, in which the items are obtained from an outside suppliers, is just a special case of the EPQ model, by letting the production rate to be almost infinite.

The EOQ model derives the optimum order size that should be placed with a vendor to minimize the holding and ordering costs. On the other hand, the EPQ model determines the optimum production size that is to be manufactured to avoid unnecessary blockage of funds and excess storage costs. Both models consider the timing of reordering or production, the cost incurred due to order/production, and the holding costs to store items; holding cost can further be in the form of rentals for the storage area, salaries of personnel looking after the inventory, electricity bills, repairs, maintenance, etc. Thus, both models describe the trade-off between fixed ordering/producing costs and variable holding costs. Furthermore, both models assume that the demand and production rates are constant over time. The traditional EOQ model assumes that the order arrives with no time, and its replenishment will happen as soon as it reaches the minimum threshold level (usually level 0). Similarly, the traditional EPQ model assumes that the production can starts immediately as the stock goes down below a minimum level (usually level 0). The price is fixed and constant while making a purchase under EOQ or producing under EPQ. The key difference between the two is that the EOQ model is applied when the items are ordered from a third party, and the EPQ model comes into use when the company is the producer itself of the products.

Unfortunately, in real-world, most of the above assumptions are unrealistic; holding and ordering costs may vary due to change in rentals, salaries of personnel, and other overhead expenses. The demand rate, as well as the price of a product, can hardly be constant. They fluctuate a lot in the real world. Consumer income, tastes, and preferences, prices of inputs and raw materials, seasonal variation in demand, etc. are key factors that will affect demand as well as price. Similarly, under the EPQ model, the production process also does not remain constant because of factors like an interruption in power supply, breakages, and repairs in plant and machinery, overheating, change in the quality of inputs and raw materials, etc.

Moreover, as a consequence of home shopping, changes in customer preferences, technological advancements, and competition, modern sales and production companies often offer a take-back guarantee. Companies soften customers' risk by offering a trial period for their products, thus, policies such as the right to return goods have become a part of daily routine. As a result, companies realize that a better understanding of returning items can provide a competitive advantage (Beltran and Krass [2], Fleischmann *et al.* [3], Pinçe *et al.* [4], Shaharudin *et al.* [5], and Barron [6, 7]).

The real need for guidance on how best to handle these uncertainties in demands, returns, productions, and costs motivates this study. We consider a continuous stochastic fluid inventory model for a single-item infinite horizon. Our main focus is to provide contributions to the study of inventory systems modulated by a Markovian environment. In the literature, dynamic control of stochastic inventory systems have been classified as periodic review models and continuous review models. In the case of continuous model, the inventory level (i.e., the number of on-hand items) can be viewed as a fluid process in which the production and demand rates undergo recurring changes in a stochastic fashion, and may be modeled as Markovian.

Markov-modulated fluid flows models have been an active area of research in recent years; one of their main applications is to the modeling the traffic evolution in communication channels. A standard example of a fluid flow is given by an infinite capacity buffer with inflow and outflow rates controlled by a Markov chain. The buffer level increases or decreases linearly at the current rate; when the buffer becomes empty, several strategies can be applied; it can remain empty until the inventory content level reaches a certain barrier (see, e.g., Boxma *et al.* [8], Kulkarni and Yan [9], Bean and O'reilly [10], and Barron and Hermel [11], Baek *et al.* [12], and Baek *et al.* [13]) or it can have positive jumps at the boundary (see e.g. Kulkarni and Yan [9, 14], and Barron [6, 15]). Fluid flow models are appropriate in situations where the arrival is comprised of a discrete unit, but the inter-arrival time between successive arrivals is negligible. Therefore, the arrival can be approximated by a continuous flow of fluid as individual units have less impact on the performance of the system. Such fluid queues are used as modeling tools of high-speed communication networks, transportation systems, congestion control systems, risk processes, and production-inventory systems.

In this study, the on-hand inventory level $\mathcal{I} = \{I(t) : t \geq 0\}$ increases or decreases according to a fluid-flow rate modulated by an n-state Continuous-Time Markov Chain (CTMC). The fluid process is the inventory position or inventory level under continuous review where the environment process represents the varying background state. A jump in the fluid level represents an external order arrival, and the transition at the background state can be the result of repairs or production facilities, etc. The cost structure includes an ordering cost for each order, a variable cost that is proportional to the actually replenished amount (both for EOQ), a set up cost for production line initialization, a production cost per item (both for EPQ), a holding cost per unit of inventory during time unit, and a penalty cost in case of shortage.

Due to the complexity of the optimal policy, these inventory/production fluid systems are challenging to optimize, and great effort in the past focused on constructing various heuristic policies (Mohebbi [16], Kouki *et al.* [17], Barron [18], and Barron and Dreyfuss [19]). Fluid versions of the EOQ model are studied in Kulkarni and Yan [9], Yan and Kulkarni [20], Kulkarni [21], Berman *et al.* [22], Berman *et al.* [23], Berman and Perry *et al.* [24], and Barron [15] and the references therein. We also mention another related model, so-called clearing system (see Kella *et al.* [25], Berman *et al.* [26], and Barron [27]), which can be regarded as a dual EOQ stochastic model. In a clearing system, the fluid process jumps back to zero when it reaches a certain positive level. For background on stochastic EPQ models, we cite Vickson [28], Kella and Whitt [29], Boxma et al. [8, 30], and Barron [7] among others.

In this chapter, our main objective is to minimize the expected discounted total cost using a discount factor $\beta > 0$. For that, we develop techniques enabling us to determine all the costs in such vendor-managing-inventory models in a closed-form. Our analysis is based on a combination of a certain martingale technique and an application of fluid flow theory to semi-regenerative processes. The martingale approach was introduced by Asmussen and Kella [31] and was frequently used in the study of inventory models (see, e.g., Boxma *et al.* [8], Kella *et al.* [25], and Barron *et al.* [27, 32]). The matrix-analytic approach and the theory of Markov-modulated fluid flows was initiated by Ramaswami [33] and Ahn *et al.* [34], who developed a unified methodology for studying a large class of insurance risk models via fluid flows by making use of the connection between the surplus process of an insurance and a particular fluid flow.

As we will show, the exit-time results are used to efficiently derive LST (Laplace–Stieltjes transform) functionals associated with the discrete-type measures, while the combination with the martingales yields simple expressions for the continuous-type measures. These explicit expressions can then be used for an analysis of the dependence of the cost functionals on the system parameters or for optimization purpose when some of these parameters (e.g. the order amount, the threshold levels, or the costs) are taken as decision variables. As such, we provide managers with a useful framework and an efficient and easy-to-implement tool to derive the best parameters and to compare the results of different demand–supply patterns.

We start by introducing the main tools of our analysis to be used.

2. Main tools

In the following, we will briefly introduce our main tools: (a) the matrix-analytic approach and the theory of Markov-modulated fluid flows, initiated by a series of papers by Ahn and Ramaswami [34–37] and Ramaswami [33] and (b) an application of the optional sampling theorem (OST) to the multi-dimensional martingale of Asmussen and Kella [31].

Here and in the following, we use bold symbols to denote vectors and blackboard to denote matrices. To be consistent, for any matrix $\mathbb{B}$, we shall denote its elements by $(\mathbb{B})_{ij}$ or by $[\mathbb{B}]_{ij}$ and reserve the notation $\mathbb{B}_{ij}$ for the sub-matrix of $\mathbb{B}$ with row indices in the set $\mathfrak{I}_i$ and column indices in the set $\mathfrak{I}_j$. Moreover, for n-vector $\mathbf{x}$, we use $\Delta_{\mathbf{x}}$ for a diagonal matrix $\Delta_{\mathbf{x}} = diag(x_1, x_2, \dots, x_n)$. We further use E and E_i to represent expectation and conditional expectation operators, respectively. $\mathbb{E}$ ($\mathbf{E}$) represents a matrix (a vector) of expectations. We denote by $\mathbf{e}$ a column vector with ones, i.e. $\mathbf{e} = (1, 1, \dots, 1)^T$, by $\mathbf{e}_i$ a row vector with the i_{th} component equal to 1 and all the other components 0, by $\mathbb{I}$ the identity matrix, by $\mathbf{0}$ the zero matrix (all with the appropriate dimensions); finally, by $\mathbf{1}_{\{A\}}$ the indicator of an event A.

2.1 Markov modulated fluid flow (MMFF)

Let $\mathfrak{I}$ be a state space that can partitioned into two sets: $\mathfrak{I} = \mathfrak{I}_1 \cup \mathfrak{I}_2$, with $|\mathfrak{I}_1| = n_1$, $|\mathfrak{I}_2| = n_2$, and $|\mathfrak{I}| = n_1 + n_2 = n$. Now, we introduce a modulating continuous time Markov chain (CTMC) $\{\mathcal{J}(t);\ t > 0\}$ with that state $\mathfrak{I}$, and a fluid process $\{\mathcal{F}(t); t \geq 0\}$ that is modulated as follows: whenever the Markov chain is $\mathcal{J}(t) = i \in \mathfrak{I}_1$, the fluid flow increases linearly at rate $c_i > 0$ and whenever it is in $\mathcal{J}(t) = j \in \mathfrak{I}_2$, the fluid flow decreases linearly at rate $|c_j| > 0$ $(c_j < 0)$. The two-dimensional stochastic process $\{\mathcal{F}(t), \mathcal{J}(t), t \geq 0\}$ is called a MMFF (Markov Modulated Fluid Flow) process. We denote by $\mathbb{Q}$ the infinitesimal generator matrix of $\mathcal{J}(t)$; $\mathbb{Q}$ is given in a block form according to transitions between the sets $\mathfrak{I}_i$ $(i = 1, 2)$. Let $\mathbb{C}_1, \mathbb{C}_2$ and $\mathbb{C}$ be diagonal matrices as follows:

$$\mathbb{C}_j = diag\{|c_i|, i \in \Im_j\}, \quad j = 1, 2.$$
$$\mathbb{C} = diag(\mathbb{C}_1, \mathbb{C}_2).$$

Let $\tau(x) = \inf(t > 0, \mathcal{F}(t) = x)$ be the first passage time to level x. Let $\Psi(\beta)$ be an $(n_1 \times n_2)$ matrix whose ij_{th} component is

$$[\Psi(\beta)]_{ij} = E\left(e^{-\beta\tau(0)}, \mathcal{J}(\tau(0)) = j | \mathcal{F}(0) = 0, \mathcal{J}(0) = i\right) \quad i \in \Im_1, j \in \Im_2,$$

which is the LST of $\tau(0)$ restricted to the event that the fluid process hits level 0 in-phase $j \in \Im_2$ and given that $\mathcal{F}(0) = 0, \mathcal{J}(0) = i \in \Im_1$. In the literature, a few algorithms, including some quadratically convergent ones, were established for computing $\Psi(\beta)$ (and all other LSTs); see, e.g., Ahn and Ramaswami [36]. Let $\tau(x,y)$ be the first passage time of $\mathcal{F}$ from level x to level y, and ${}_a^b\tau(x,y)$ be the first passage time of $\mathcal{F}$ from level x to level y avoiding a visit below a or above b (for simplicity, we use ${}^b\tau(x,y) \equiv {}_0^b\tau(x,y)$ and ${}_a\tau(x,y) \equiv {}_a^\infty\tau(x,y)$). Let $\hat{f}(x,y,\beta)$ and ${}_a^b\hat{f}(x,y,\beta)$ denote, respectively, the LST matrices of the joint distribution of the first passage times $\tau(x,y)$, ${}_a^b\tau(x,y)$ and the state of the phase process at each first passage time.

An important variant of the fluid flow $\mathcal{F}$, a *Reflected Fluid Flow,* is particularly useful in the analysis of our inventory level process. The reflected fluid flow $\mathcal{F}^r$ is obtained by reversing the roles of the up and down environment states. Analogous, $\Psi^r(\beta)$ is the matrix (of order $(n_2 \times n_1)$) whose (i,j) component is the LST of the time to reach the level 0 for the process $\mathcal{F}^r$ restricted to $\mathcal{J}^r(\tau(0)) = j \in \Im_1$, given that $\mathcal{F}^r(0) = 0$ and $\mathcal{J}^r(0) = i \in \Im_2$, where $\mathcal{J}^r(t)$ is the modulated state process for $\mathcal{F}^r$ (we use notations $\hat{f}^r(x,y,\beta)$ and ${}_a^b\hat{f}^r(x,y,\beta)$ to denote quantities similar to those above defined for $\mathcal{F}^r$).

All these matrices, for hitting times, that we will use are straightforward to evaluate once we have computed $\Psi(\beta)$. **Table 1** displays the basic elements (matrices) for the derivation of these LSTs; the first three matrices are associated with to flow $\mathcal{F}$, while the next three matrices are associated to the rate-reverse flow $\mathcal{F}^r$ by interchanging the indices 1 and 2. The LST matrices and their sizes are given in **Table 2**; all matrices have nice probabilistic interpretations. For more details see Ramaswami [33], Ahn *et al.* [34], and Bean *et al.* [38].

Quantity	Matrix size	The process
$\mathbb{K}(\beta) = \mathbb{C}_1^{-1}(\mathbb{Q}_{11} - \beta\mathbb{I}) + \Psi(\beta)\mathbb{C}_2^{-1}\mathbb{Q}_{12}$	$n_1 \times n_1$	$\mathcal{F}$
$\mathbb{H}(\beta) = \mathbb{C}_2^{-1}(\mathbb{Q}_{22} - \beta\mathbb{I} + \mathbb{Q}_{21}\Psi(\beta))$	$n_2 \times n_2$	$\mathcal{F}$
$\Xi(\beta,x) = \Psi(\beta)\int_{(0,x)} e^{\mathbb{H}(\beta)y}(\mathbb{C}_2^{-1}\mathbb{Q}_{21})e^{\mathbb{K}(\beta)y}dy$	$n_1 \times n_1$	$\mathcal{F}$
$\mathbb{K}^r(\beta) = \mathbb{C}_2^{-1}\mathbb{Q}_{22} + \Psi^r(\beta)\mathbb{C}_1^{-1}\mathbb{Q}_{12}$	$n_2 \times n_2$	$\mathcal{F}^r$
$\mathbb{H}^r(\beta) = \mathbb{C}_1^{-1}(\mathbb{Q}_{11} - \beta\mathbb{I} + \mathbb{Q}_{12}\Psi^r(\beta))$	$n_1 \times n_1$	$\mathcal{F}^r$
$\Xi^r(\beta,x) = \Psi^r(\beta)\int_{(0,x)} e^{\mathbb{H}^r(\beta)y}(\mathbb{C}_1^{-1}\mathbb{Q}_{12})e^{\mathbb{K}^r(\beta)y}dy$	$n_2 \times n_2$	$\mathcal{F}^r$

Table 1.
Transform matrices.

LST	The LST of the First Passage Time	Matrix size
${}_0\hat{f}_{12}(x,0,\beta) = \Psi(\beta)e^{\mathbb{H}(\beta)x}$	*from* $(x,\Im_1)$ *to* $(0,\Im_2)$ *avoiding* 0 *in* F	$n_1 \times n_2$
${}_0\hat{f}_{22}(x,0,\beta) = e^{\mathbb{H}(\beta)x}$	*from* $(x,\Im_2)$ *to* $(0,\Im_2)$ *avoiding* 0 *in* F	$n_2 \times n_2$
${}_0\hat{f}_{11}(0,x,\beta) = e^{\mathbb{K}(\beta)x}(\mathbb{I}+\Xi(\beta,x))^{-1}$	*from* $(0,\Im_1)$ *to* $(x,\Im_1)$ *avoiding* 0 *in* F	$n_1 \times n_1$
${}^x\Psi^r(\beta) = \Psi^r(\beta) - {}_0\hat{f}^r_{22}(0,x,\beta)\Psi^r(\beta)e^{\mathbb{H}^r(\beta)x}$	*from* $(0,\Im_2)$ *to* $(0,\Im_1)$ *avoiding* x *in* F^r	$n_2 \times n_1$
${}_0\hat{f}^r_{22}(0,x,\beta) = e^{\mathbb{K}^r(\beta)x}(\mathbb{I}+\Xi^r(\beta,x))^{-1}$	*from* $(0,\Im_2)$ *to* $(x,\Im_2)$ *avoiding* 0 *in* F^r	$n_2 \times n_2$
$\hat{f}_{11}(0,x,\beta) = \left[\mathbb{I} - {}^x\Psi(\beta)(\beta\mathbb{I}-\mathbb{Q}_{22})^{-1}\mathbb{Q}_{21}\right]^{-1} \times {}_0\hat{f}_{11}(0,x,\beta)$	*from* $(0,\Im_1)$ *to* $(x,\Im_1)$ *in* F	$n_1 \times n_1$
$\hat{f}_{21}(0,x,\beta) = (\beta\mathbb{I}-\mathbb{Q}_{22})^{-1}\mathbb{Q}_{21}\hat{f}_{11}(0,x,\beta).$	*from* $(0,\Im_2)$ *to* $(x,\Im_1)$ *in* F	$n_2 \times n_1$

Table 2.
LST of first passage times.

2.2 The multi-dimensional martingale

Let $\{\mathcal{X}(t), t \geq 0\}$ be a right continuous Markov modulated Lévy process with modulating process $\{\mathcal{J}(t), t \geq 0\}$ which is a right continuous irreducible finite state space continuous time Markov chain. Let $\{Y(t), t \geq 0\}$ be an adapted continuous process with a finite expected variation on finite intervals and let $\mathcal{Z}(t) = \mathcal{X}(t) + Y(t)$. Asmussen and Kella [31] have shown, that for such a process, the matrix with elements $E_i\left[e^{\alpha\mathcal{X}(t)}; \mathcal{J}(t) = j\right]$ has the form of $e^{t\mathbb{K}(\alpha)}$ for some matrix $\mathbb{K}(\alpha)$. Theorem 2.1 in Asmussen and Kella [31] yields that under certain conditions on $\{\mathcal{Z}(t), t \geq 0\}$, the multi-dimensional process

$$M(\alpha,t) = \int_{v=0}^{t} e^{\alpha\mathcal{Z}(v)}\mathbf{1}_{\mathcal{J}(v)}dv\mathbb{K}(\alpha) + e^{\alpha\mathcal{Z}(0)}\mathbf{1}_{\mathcal{J}(0)} - e^{\alpha\mathcal{Z}(t)}\mathbf{1}_{\mathcal{J}(t)} + \alpha\int_{v=0}^{t} e^{\alpha\mathcal{Z}(v)}\mathbf{1}_{\mathcal{J}(v)}dY(v) \tag{1}$$

is a (row) vector-valued zero mean martingale. Some of the relevant functionals in this paper will be obtained by applying the OST (or Doob's optional sampling theorem, see Doob [39]) to appropriate special cases of For our models, the inventory level $\mathcal{I}$ is a special case of $\mathcal{X}(t)$ and has piecewise linear sample paths, with slope c_j on intervals where $\mathcal{J}(t) = j$.

The outline of the chapter is as follows. We start with the ordering model and introduce an extension to the EOQ model, a *fluid EOQ-type* inventory model. Then, using the *fluid EOQ-type* model as a key, our study is generalized to include production facility, the *fluid EPQ-type* model.

3. A fluid EOQ-type inventory model

The simple *Economic Ordering Quantity* (EOQ) model assumes an inventory model of infinite horizon for a single item. The demand occurs at a constant rate and each time the inventory level hits a level 0, a fixed and immediate order of size Q is places.

EOQ considers the timing of reordering, the cost incurred to place an order, and the costs to store merchandise. Here, we generalize the traditional EOQ model and consider demand and return rates that depend on the background environment, and varying state-dependent holding and ordering costs. We start with the mathematical description of the model.

Let $\mathcal{I} = \{I(t) : t \geq 0\}$ be the on-hand inventory level at time t. The rate of change of the level is modulated by a continuous time Markov chain (CTMC) $\{\mathcal{J}(t) : t \geq 0\}$ on a finite state space $\mathfrak{I} = \{1, 2, \dots, n\}$ with a generator matrix $\mathbb{Q} = \left[Q_{ij}\right]$. As long as $\mathcal{J}(t)$ is in state i, there is a demand at rate d_i and a return at rate r_i. The net rate is thus $c_i = r_i - d_i$. Note that c_i may be either negative or positive. Accordingly, we have two disjoint sets $(\mathfrak{I}_1, \mathfrak{I}_2)$, $\mathfrak{I} = \mathfrak{I}_1 \cup \mathfrak{I}_2$, where $\mathfrak{I}_1$ is a non-empty set of increasing rates $\mathfrak{I}_1 = \{i \in \mathfrak{I} : c_i > 0\}$ and $\mathfrak{I}_2$ is a non-empty set of the decreasing rates $\mathfrak{I}_2 = \{i \in \mathfrak{I} : c_i < 0\}$. Let $|\mathfrak{I}_1| = n_1$ and $|\mathfrak{I}_2| = n_2$; thus $n_1 + n_2 = n$. Let $\boldsymbol{\pi} = [\pi_1, \dots, \pi_n]$ be the limiting distribution of the $\mathcal{J}(t)$ process, i.e. $\boldsymbol{\pi}$ is the unique solution to

$$\boldsymbol{\pi}\mathbb{Q} = 0, \boldsymbol{\pi}\mathbf{e} = 1.$$

The system is stable if and only if the expected input rate is negative, i.e.,

$$\sum_{i=1}^{n} \pi_i c_i < 0.$$

When $I(t)$ down-crosses level 0, an order of size $Q > 0$ from an external supplier is placed which arrives instantaneously. Thus, immediately after the down-crossing to emptiness by the inventory level in state i, the process restarts at level Q. It should be noted, that the EOQ model is a special case of the base–stock polices, in particular, the (s, S)-type, where the reorder level $s = 0$, the replenishment up-to-level $S = Q$ and zero lead time.

Let T_k be the time of the k_{th} jump ($T_0 = 0$) and $\mathcal{J}_k = \mathcal{J}(T_k)$ be the environmental state at T_k (just after the jump). Note that since an order is placed when the inventory drops to level 0, the state $\mathcal{J}_k$ has to be a descending one, i.e., $\mathcal{J}_k \in \mathfrak{I}_2$. We assume that over $[T_{k-1}, T_k)$ $k = 1, 2..$, the process $\{\mathcal{J}(t), t \in T_{k-1}, T_k)\}$ is an irreducible CTMC (continuous time Markov chain) on $\mathfrak{I}$. We call the points where the process jumps up (the replenishment times) *order points*. They form a semi-renewal process where $T_k's$ are the semi regenerative points of the process (see Ross [40]). Define the k_{th} cycle as the time elapsed between T_{k-1} and $T_k, k = 1, 2, \dots$ Denote by $C_k = T_k - T_{k-1}$, $k = 1, 2 \dots$ $(C_0 = 0)$ the inter-replenishment times; let $C = C_1$. **Figure 1** illustrates a sample path of the process $I(t), t \geq 0$; for simplicity, we assume that $I(0) = Q, \mathcal{J}(0) \in \mathfrak{I}_n$ and let $\gamma = \left[\gamma_1, \gamma_2, \dots, \gamma_{n_2}\right]$ be the initial probability vector of $\mathcal{J}(0)$.

3.1 The cost structure

Our cost structure composed of two components: (a) an order cost, including a fixed cost whenever an order is placed and a purchasing cost; (b) a holding cost for the stock. We further assume that the cost rates are determined by the environment at the *order points;* in real-world, order rates usually depend on the state of the economy, the season, and change accordingly. Specifically, if the state at an *order point* is $\mathcal{J}_k = i \in \mathfrak{I}_2$, the fixed ordering cost is K_i for an order (typically cost of ordering and shipping and handling), the cost to purchase one item from an external supplier is q_i,

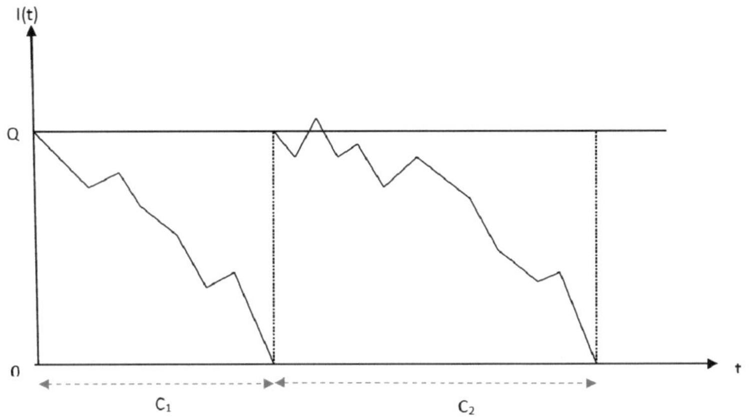

Figure 1.
A typical path of the fluid EOQ-type inventory process.

and the cost to hold one item in inventory during a time interval of length dt is $h_i dt$ during all the cycle k in which $\mathcal{J}_k = i \in \mathfrak{I}_2$ (meaning that the holding cost for one unit is constant between two consecutive order replacements, i.e. $h(t) = h_i$ for $t \in T_k, T_{k+1})$ and $\mathcal{J}_k = i \in \mathfrak{I}_2$). We next introduce the functionals indicating the expected discounted costs using a discount factor $\beta > 0$.

a. *Order cost*. If $\mathcal{J}_k = i \in \mathfrak{I}_2$, an order of size Q is immediately replenished up to level $I(T_k) = Q$. The order cost is $K_i + q_i Q$. Let $OC(\beta)$ be the expected discounted order cost and let $\hat{\mathbf{O}}(\beta)$ be an $(n_2 \times 1)$ vector whose i_{th} component $\hat{O}_i(\beta)$ is given by

$$\hat{O}_i(\beta) = E_i\left[\sum_{k=0}^{\infty} e^{-\beta T_k}\left(K_i + q_i Q\right)\right], \tag{2}$$

i.e., the *expected* discounted order cost, given $\mathcal{J}(0) = i \in \mathfrak{I}_2, I(0) = Q$. Then we have $OC(\beta) = \gamma \cdot \hat{\mathbf{O}}(\beta)$. Applying **Table 2**, let ${}_0\hat{f}_{22}(Q, 0, \beta)$ be an $(n_2 \times n_2)$ matrix whose ijth components is

$$\left({}_0\hat{f}_{22}(Q, 0, \beta)\right)_{ij} = e^{\mathbb{H}(\beta)Q}, \quad i \in \mathfrak{I}_2, j \in \mathfrak{I}_2. \tag{3}$$

The *component* $\left({}_0\hat{f}_{22}(Q, 0, \beta)\right)_{ij}$ represents the LST of the time until the process hits level 0 in state $j \in \mathfrak{I}_2$, given $\mathcal{J}(0) = i \in \mathfrak{I}_2, I(0) = Q$ (for a proof, see Ramaswami [33], Theorem 5).

Lemma 3.1 *The total expected discounted order cost vector* $\hat{\mathbf{O}}(\beta)$ *of order* $(n_2 \times 1)$ *satisfies the following equation:*

$$\hat{\mathbf{O}}(\beta) = \left(\mathbb{I} - {}_0\hat{f}_{22}(Q, 0, \beta)\right)^{-1} \Delta_{\mathbf{K}+\mathbf{q}Q}\, \mathbf{e}. \tag{4}$$

where $\Delta_{\mathbf{K}+\mathbf{q}Q} = diag\left(K_i + q_i Q\right), i \in \mathfrak{I}_2$.

Proof. It is easy to verify that $\hat{\mathbf{O}}(\beta)$ can be written as (recall that C is the time of the next order):

$$\hat{\mathbf{O}}(\beta) = \Delta_{\mathbf{K}+\mathbf{q}Q}\,\mathbf{e} + \mathbb{E}\left(e^{-\beta C}\right)\hat{\mathbf{O}}(\beta), \tag{5}$$

where the $(n_2 \times n_2)$ matrix $\mathbb{E}\left(e^{-\beta C}\right)$ is the LST of the cycle length; its ij_{th} component is given by

$$\mathbb{E}\left(e^{-\beta C}\right)_{ij} = E\left(e^{-\beta C}\mathbf{1}_{\{\text{level } 0 \text{ hit at time } C \text{ in phase } j\}} \middle| \mathcal{J}(0) = i, I(0) = Q\right). \tag{6}$$

Applying the fluid model $\mathbb{E}\left(e^{-\beta C}\right) = {}_0\hat{f}_{22}(Q, 0, \beta)$ and solving (5) for $\hat{\mathbf{O}}(\beta)$ we obtain (4) ■

a. *Holding cost.* The expected discounted holding cost can be expressed as

$$HC(\beta) = E\left(\int_0^\infty h(t)e^{-\beta t}I(t)dt\right). \tag{7}$$

Let $\hat{\mathbf{H}}(\beta)$ be an $(n_2 \times 1)$ vector whose i_{th} component is given by

$$\hat{H}_i(\beta) = E_i\left(\int_0^\infty h(t)e^{-\beta t}I(t)dt\right) \tag{8}$$

i.e., the expected discounted holding cost, given $\mathcal{J}(0) = i \in \mathfrak{I}_2$. Thus, we have $HC(\beta) = \gamma \cdot \hat{\mathbf{H}}(\beta)$.

Lemma 3.2 *The vector* $\hat{\mathbf{H}}(\beta)$ *of order* $(n_2 \times 1)$ *satisfies the following equation:*

$$\hat{\mathbf{H}}(\beta) = \left(\mathbb{I} - {}_0\hat{f}_{22}(Q, 0, \beta)\right)^{-1}\Delta_{\mathbf{h}}\mathbf{E}\left(\int_{t=0}^{C} e^{-\beta t}I(t)dt\right).$$

Proof. Applying the regenerative theory, and similar to Lemma 3.1, the vector $\hat{\mathbf{H}}(\beta)$ can be written as

$$\hat{\mathbf{H}}(\beta) = \Delta_{\mathbf{h}}\mathbf{E}\left(\int_{t=0}^{C} e^{-\beta t}I(t)dt\right) + \mathbb{E}\left(e^{-\beta C}\right)\hat{\mathbf{H}}(\beta)$$

(The first vector $\mathbf{E}\left(\int_{t=0}^{C} e^{-\beta t}I(t)dt\right)$ is the expected discounted inventory level of the first cycle). Now, we use the OST to the multi-dimensional martingale to find the $(n_2 \times 1)$ vector $\mathbf{E}\left(\int_{t=0}^{C} e^{-\beta t}I(t)dt\right)$. For $\mathcal{J}(0) = i \in \mathfrak{I}_2, I(0) = Q,$ consider a Lévy process $\{\mathcal{X}_i(t)\}$ as follows:

$$\mathcal{X}_i(t) = \mathcal{X}_i(0) - \int_{v=0}^{t} c_{\mathcal{J}(v)}dv \quad 0 \le t < C, \quad \mathcal{X}_i(0) = -Q. \tag{9}$$

It is not difficult to see that the latter process up to time C, i.e., $(\mathcal{X}_i(t))_{0\le t<C}$, has the same distribution as $(-I(t))_{0\le t<C}$. Chapter XI, p. 311 of Asmussen [41] yields that

$$E_i\left[e^{\alpha\mathcal{X}_i(t)};\mathcal{J}(t)=j\right]=\left(e^{t\mathbb{K}(\alpha)}\right)_{ij}$$

where

$$\mathbb{K}(\alpha)=\mathbb{Q}-\alpha\mathbb{C}. \tag{10}$$

Let $Y(t)=-(\beta/\alpha)t$ (for an arbitrary $\alpha>0$) and let $\mathcal{Z}_i(t)=\mathcal{X}_i(t)+Y(t)$. Since $Y(t)$ is adapted and has paths of a finite expected variation, the process

$$\begin{aligned}M_i(\alpha,t)&=\int_{v=0}^{t}e^{\alpha\mathcal{Z}_i(v)}\mathbf{1}_{\mathcal{J}(v)}dv\mathbb{K}(\alpha)+e^{\alpha\mathcal{Z}_i(0)}\mathbf{1}_{\mathcal{J}(0)}-e^{\alpha\mathcal{Z}_i(t)}\mathbf{1}_{\mathcal{J}(t)}+\alpha\int_{v=0}^{t}e^{\alpha\mathcal{Z}_i(v)}\mathbf{1}_{\mathcal{J}(v)}dY(v)\\&=\int_{v=0}^{t}e^{\alpha\mathcal{X}_i(v)-\beta v}\mathbf{1}_{\mathcal{J}(v)}dv(\mathbb{K}(\alpha)-\beta\mathbb{I})+e^{\alpha\mathcal{X}_i(0)}\mathbf{1}_{\mathcal{J}(0)}-e^{\alpha\mathcal{X}_i(t)-\beta t}\mathbf{1}_{\mathcal{J}(t)}\end{aligned} \tag{11}$$

is an n_2-dimensional row vector-valued zero mean martingale. The OST yields $EM_i(\alpha,0)=EM_i(\alpha,C)=0$, i.e.,

$$E_i\left(\int_{t=0}^{C}e^{\alpha X_i(t)-\beta t}dt\right)=\left[\mathbb{E}_i\left(e^{\alpha\mathcal{X}_i(C)-\beta C}\mathbf{1}_{\mathcal{J}(C)}\right)-\mathbb{E}_i\left(e^{\alpha\mathcal{X}_i(0)}\mathbf{1}_{\mathcal{J}(0)}\right)\right](\mathbb{K}(\alpha)-\beta\mathbb{I})^{-1}\mathbf{e}. \tag{12}$$

Obviously, we have $E_i\left(e^{\alpha\mathcal{X}_i(0)}\mathbf{1}_{\mathcal{J}(0)}\right)=e^{-\alpha Q}$ **for** $i\in\mathfrak{I}_2$, or in an $(n_2\times n)$ matrix form

$$\mathbb{E}\left(e^{\alpha\mathcal{X}(0)}\mathbf{1}_{\mathcal{J}(0)}\right)=\left(\mathbf{0}\;e^{-\alpha Q}\mathbb{I}\right). \tag{13}$$

(Here, the $(n_2\times n_1)$ zero matrix arising due to the fact that $\mathcal{J}(0)\in\mathfrak{I}_2$). Next, we have to derive $\mathbf{E}_i\left(e^{\alpha X_i(C)-\beta C}\mathbf{1}_{\mathcal{J}(C)}\right)$. Since $X_i(C)=0$, the fluid method yields the $(n_2\times n)$ matrix form

$$\mathbb{E}\left(e^{-\beta C}\mathbf{1}_{\mathcal{J}(C)}\right)=\left(\mathbf{0}_0\hat{f}_{22}(Q,0,\beta)\right). \tag{14}$$

Substituting (13) and (14) in (12) we obtain the $(n_2\times 1)$ vector

$$\mathbf{E}\left(\int_{t=0}^{C}e^{\alpha\mathcal{X}(t)-\beta t}dt\right)=\left(\left(\mathbf{0}_0\hat{f}_{22}(Q,0,\beta)\right)-\left(\mathbf{0}\;e^{-\alpha Q}\mathbb{I}\right)\right)(\mathbb{K}(\alpha)-\beta\mathbb{I})^{-1}\mathbf{e}. \tag{15}$$

Finally, take the derivative of both sides of (15) with respect to α, let $\alpha=0$ and note that

$$\mathbf{E}\left(\int_{t=0}^{C} e^{-\beta t}\mathcal{X}(t)dt\right) = -\mathbf{E}\left(\int_{t=0}^{C} e^{-\beta t}I(t)dt\right)$$

leads to

$$\mathbf{E}\left(\int_{t=0}^{C} e^{-\beta t}I(t)dt\right) = -\frac{d}{d\alpha}\left[\left(\mathbf{0}_0\hat{f}_{22}(Q,0,\beta)\right) - \left(\mathbf{0}\ e^{-\alpha Q}\mathbb{I}\right)(\mathbb{K}(\alpha)-\beta\mathbb{I})^{-1}\right]\mathbf{e}\Big|_{\alpha=0}. \quad (16)$$

Accordingly, the expected discounted total cost is

$$TC(\beta) = OC(\beta) + HC(\beta). \quad (17)$$

■

Remark 1 *It is easy to extend this model to include an order size determined by the environment at the order point, i.e., if $\mathcal{J}(T_k) = i \in \mathfrak{I}_2$, an order of Q_i items is placed (and, accordingly, $I(0) = Q_i$ with probability $\gamma_i, i \in \mathfrak{I}_2$). For that we define the vector $\mathbf{Q} = (Q_1, \ldots, Q_{n_2})$. Now, the entries $\left({}_0\hat{f}_{22}(Q,0,\beta)\right)_{ij}$ are obtained by replacing Q with Q_i in* (3). *The order cost is given by* (4) *with $\Delta_{\mathbf{K}+\mathbf{q}\mathbf{Q}}$ replacing $\Delta_{\mathbf{K}+\mathbf{q}Q}$, and the holding cost is given by substituting $\Delta_{e^{-\alpha\mathbf{Q}}}$ instead $e^{-\alpha Q}\mathbb{I}$ in* (13) *(here, $\mathcal{X}_i(0) = -Q_i$).*

Using the approach for the *fluid EOQ-type* model as a key, we next include a production facility and introduce the *fluid EPQ-type* model.

4. A fluid EPQ-type inventory model

Similar as above, let $I(t)$ be the inventory level at a time t. The process $I(t)$ can be partitioned into two parts, $I^+(t)$ and $I^-(t)$. The first part of the cycle, $I^+(t)$, is the ON period and this period ends whenever $I^+(t)$ reaches a predetermined level Q. The second part of the cycle, $I^-(t)$, is the OFF period and this is the time until $I^-(t)$ drops to level 0. The ON period is characterized by stochastic inputs, production and returns, and outputs, demand. However, no production occurs during the OFF period, and thus, the OFF period is characterized by stochastic returns and demand. Here, the analysis is more challenging due to the periods ON and OFF. Denote by $\{p_1, \ldots, p_n\} \subset (0,\infty)$ the production rates, by $\{d_1, \ldots, d_n\} \subset (0,\infty)$ the demand rates, and by $\{r_1, \ldots, r_n\} \subset (0,\infty)$ the returns rates. The rate at which the inventory is filled at time t is determined by the current environmental state $\mathcal{J}(t)$. During the ON period (OFF period) and as long as $\mathcal{J}(t) = i$, the growth rate is the difference of the production and return rates (only the return rate) and the demand rate, $c_i^+ = r_i + p_i - d_i$ ($c_i^- = r_i - d_i$); similarly as before, $c_i, i = 1, .., n$ may be either negative or positive. We do not allow backlog; thus when $I^+(t)$ drops to level 0 (due to high demands), it stays there as long as the environmental growth rate is negative and until the environmental state changes to some positive growth rate. Note that the behavior of the process during each period is different. During the OFF period, we enforce that $\sum_{i=1}^{n}\pi_i c_i^- < 0$ (a necessary and sufficient condition for the stability of the OFF process). During the ON period, we assume that $\sum_{i=1}^{n}\pi_i c_i^+ > 0$ (although with the absence of this, $I^+(t)$ is stable due to the

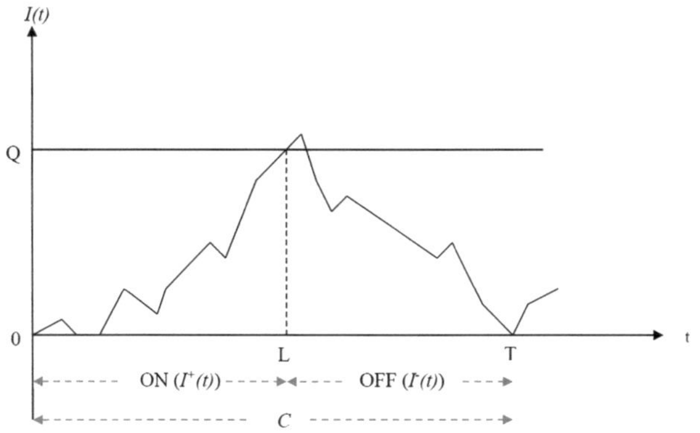

Figure 2.
A typical path of the fluid EPQ-type inventory process.

reflection at level 0, however, it is a realistic assumption in order to avoid high lost demand).

A typical sample path of the inventory process is given in **Figure 2**. As we see, the inventory process is a semi-regenerative process which alternates between ON periods and OFF periods. We further note that if the ON periods are deleted from the sample path and the OFF periods are glued together we obtain a *fluid EOQ-type* model with refilling every time level 0 is reached. This holds with one exception; while each cycle under the *fluid EOQ-type* process starts with a descending state, it's not necessarily holds in the *fluid EPQ-type* model (a detailed explanation is given below).

Define the following stopping times:

$$T_0 = 0$$
$$T_k = \inf\{t > L_k : I(t) = 0\} \ k = 1, 2 \ldots$$
$$L_k = \inf\{t \geq T_{k-1} : I(t) = Q\} k = 1, 2, \ldots$$

$T_k, k = 1, 2, \ldots$ are the times of switchings from OFF to ON, and $L_n, n = 1, 2, \ldots$ are the time instants of switchings from ON to OFF. Thus, $I(t)$ is a semi-regenerative process with $T'_n s$ are semi-regenerative points. Define the k_{th} cycle as the time elapsed between T_{k-1} and $T_k, k = 1, 2, ..$ and let $C_k = T_k - T_{k-1}, k = 1, 2 \ldots$ be the kth cycle length. We use the generic form $L = L_1, T = T_1, T' = T - L$ and $C = C_1$ (so $C = L + T'$). Note that conditioning on the state at time L and the common background environmental process, the two process $I^+(t)_{0 \leq t \leq L}$ (the ON period) and $I^-(t)_{L \leq t \leq T}$ (the OFF period) are independent.

We construct diagonal matrices $\mathbb{C}_j^+ = diag\{|c_i^+|, \ i \in \mathfrak{I}_j\}, \mathbb{C}_j^- = diag\{|c_i^-|,$ $i \in \mathfrak{I}_j\}, j = 1, 2$ and $\mathbb{C}^+ = diag(\mathbb{C}_1^+, \mathbb{C}_2^+)$ and $\mathbb{C}^- = diag(\mathbb{C}_1^-, \mathbb{C}_2^-)$ from these rates. Regarding the fluid EPQ-type model, each ON/OFF period has one type of rates ($\mathbb{C}^+$ or $\mathbb{C}^-$); hence, given the state at switching epoch and the common environment, the ON/OFF periods are independent. Now, we can analyze the inventory level within each period independently using MMFF process. Specifically, we consider I^+, $\mathbb{C}^+$ and $\Psi^+(\beta)$ (I^-, $\mathbb{C}^-$, $\Psi^-(\beta)$) corresponding to the ON (OFF) period. Note that, for this model, each LST matrix should be derived for ON (marked as +) and OFF (marked as −) processes. However, we do not insert the marks + or − corresponding to ON/OFF processes; it should be clear from the context which of the marks applies.

4.1 The cost functionals

We consider four costs: (a) the setup cost, (b) the holding cost of the inventory, (c) the production cost, and (d) the unsatisfied demand cost. For the derivation of the costs, we first need to derive the expected discounted cycle length $(n \times n)$ matrix $\mathbb{E}(e^{-\beta C})$.

4.2 The matrix $\mathbb{E}(e^{-\beta C})$

Given the state $\mathcal{J}(L)$, the ON period and the OFF period are independent. Clearly, the ON period ends at state in $\mathfrak{J}_1$ (at time L). However, at that point of time, the production stops, and thus, the state at which the OFF period starts can be either in $\mathfrak{J}_1$ or in $\mathfrak{J}_2$. Similarly, the OFF period ends at state in $\mathfrak{J}_2$ (at time T). However, since the production starts at that point, the ON period can start either in state in $\mathfrak{J}_1$ or in $\mathfrak{J}_2$. This needs to be considered, particularly at switching times. Hence, at these switching times, the phases are labeled by arranging the entries according to the new states. Given the state $\mathcal{J}(L)$, and due to that independency, the matrix $\mathbb{E}(e^{-\beta C})$ is given by

$$\mathbb{E}(e^{-\beta C}) = \mathbb{E}(e^{-\beta L})\mathbb{E}\left(e^{-\beta T'}\right). \tag{18}$$

We first introduce two similar LST matrices which differ only with their initial environment: (i) an $(n_1 \times n_1)$ matrix $\hat{f}_{11}(0, Q, \beta)$ whose ijth component $\left(\hat{f}_{11}(0, Q, \beta)\right)_{ij}$ represents the LST of the time until hitting Q in state $j \in \mathfrak{J}_1$, given $I(0) = 0, \mathcal{J}(0) = i \in \mathfrak{J}_1$ and (ii) an $(n_2 \times n_1)$ matrix $\hat{f}_{21}(0, Q, \beta)$ whose ijth component $\left(\hat{f}_{21}(0, Q, \beta)\right)_{ij}$ represents the LST of the time until hitting Q in state $j \in \mathfrak{J}_1$, given $I(0) = 0, \mathcal{J}(0) = i \in \mathfrak{J}_2$. Now, define the $(n \times n_1)$ matrix:

$$\hat{f}_{\cdot 1}(0, Q, \beta) = \begin{pmatrix} \hat{f}_{11}(0, Q, \beta) \\ \hat{f}_{21}(0, Q, \beta) \end{pmatrix}.$$

Since the ON period ends with environment in $\mathfrak{J}_1$ (at time L), the $(n \times n)$ matrix $\mathbb{E}(e^{-\beta L})$ has the form:

$$\mathbb{E}(e^{-\beta L}) = \left(\hat{f}_{\cdot 1}(0, Q, \beta) \;\; \mathbf{0}\right). \tag{19}$$

Similarly, define the $(n \times n_2)$ matrix:

$${}_0\hat{f}_{\cdot 2}(Q, 0, \beta) = \begin{pmatrix} {}_0\hat{f}_{12}(Q, 0, \beta) \\ {}_0\hat{f}_{22}(Q, 0, \beta) \end{pmatrix}.$$

(See also (3)). The OFF period ends with environment in $\mathfrak{J}_2$ (at time T), and thus, the $(n \times n)$ matrix $\mathbb{E}(e^{-\beta T'})$ has the form of

$$\mathbb{E}\left(e^{-\beta T'}\right) = \left(\mathbf{0}\, {}_0\hat{f}_{\cdot 2}(Q, 0, \beta)\right). \tag{20}$$

(Recall that all matrices are given in **Table 1**). Next, we derive the expected discounted costs.

a. **Set up cost**. Let K_1 be the setup cost to switch from OFF to ON (at time T) and K_2 be the setup cost to switch from ON to OFF (at time L). Let $SC(\beta)$ be the expected discounted set up cost and let $\hat{\mathbf{S}}(\beta)$ be an $(n \times 1)$ vector whose i_{th} component is the expected discounted set up cost given $\mathcal{J}(0) = i \in \Im, I(0) = 0$,

$$\hat{\mathbf{S}}(\beta) = E_i \sum_{n=1}^{\infty} (K_2 \exp(-\beta L_n) + K_1 \exp(-\beta T_n)). \tag{21}$$

Similar technique as before arrives at

$$\begin{aligned} SC(\beta) &= \gamma \hat{\mathbf{S}}(\beta) \\ \hat{\mathbf{S}}(\beta) &= \left(\mathbb{I} - \mathbb{E}\left(e^{-\beta C}\right)\right)^{-1} \left(K_1 \mathbb{E}\left(e^{-\beta C}\right) + K_2 \mathbb{E}\left(e^{-\beta L}\right)\right)\mathbf{e}. \end{aligned} \tag{22}$$

Substituting (18) and (19) finalizes the derivation.

b. **Holding cost.** The total expected discounted holding cost $HC(\beta) = \gamma \hat{\mathbf{H}}(\beta)$, where the vector $\hat{\mathbf{H}}(\beta)$ is of order $(n \times 1)$; its i_{th} components is the discounted expected holding cost given $\mathcal{J}(0) = i \in \Im, I(0) = 0$. Revoking the ergodic theorem for regenerative process, we can write $\hat{\mathbf{H}}(\beta)$ in terms of the first cycle and have

$$\hat{\mathbf{H}}(\beta) = \left(\mathbb{I} - \mathbb{E}\left(e^{-\beta C}\right)\right)^{-1} \Delta_{\mathbf{h}} \left[\mathbf{E}\left(\int_0^L e^{-\beta t} I^+(t) dt\right) + \mathbb{E}\left(e^{-\beta L}\right) \mathbf{E}\left(\int_0^{T'} e^{-\beta t} I^-(t+L) dt\right)\right]. \tag{23}$$

The basic tool we use to derive $\mathbf{E}\left(\int_0^L e^{-\beta t} I^+(t) dt\right)$ and $\mathbf{E}\left(\int_0^{T'} e^{-\beta t} I^-(t+L) dt\right)$ is OST to the Asmussen–Kella multi-dimensional martingale, as introduced in Lemma 3.2. Here, we consider the process

$$\mathcal{X}(t) = \int_{v=0}^{t} c_{\mathcal{J}(v)} dv \quad 0 \le t \le L, \mathcal{X}(0) = 0. \tag{24}$$

Chapter *XI*, p. 311 of Asmussen [41] yields that $E_i\left[e^{\alpha \mathcal{X}(t)}; \mathcal{J}(t) = j\right] = \left(e^{t\mathbb{K}(\alpha)}\right)_{ij}$ where $\mathbb{K}(\alpha) = \mathbb{Q} + \alpha \mathbb{C}$ (specifically, $\mathbb{K}^+(\alpha) = \mathbb{Q} + \alpha \mathbb{C}^+$). let $\mathcal{L}(t) = -\min_{0 \le v \le t} \mathcal{X}(v)$. The process $\mathcal{L}(t)$, known as the *local time,* is a non-decreasing process that increases only whenever $I^+(t) = 0$ (for more details on *local time* and its properties, we refer the interested reader to [41]). Next, we let $\mathcal{Z}(t) = \mathcal{X}(t) + \mathcal{L}(t)$. It is not difficult to see that the latter process up to time L, i.e., $(\mathcal{Z}(t))_{0 \le t < L}$, has the same distribution as $(I(t))_{0 \le t < L}$. Finally, define $Y(t) = \mathcal{L}(t) - (\beta/\alpha)t$, for arbitrary $\beta \ge 0$ and $\alpha < 0$, and $W(t) = \mathcal{X}(t) + Y(t) = \mathcal{Z}(t) - (\beta/\alpha)t$. Since Y is adapted and has paths of finite expected total variation on bounded intervals, Theorem 2.1 of Asmussen and Kella [31] leads to the next claim.

Claim 4.1 *The* $(n \times 1)$ *vector* $\mathbf{E}\left(\int_0^L e^{-\beta t} I^+(t)dt\right)$ *is given by:*

$$\mathbf{E}\left(\int_0^L e^{-\beta t} I^+(t)dt\right) = \frac{d}{d\alpha}\mathbf{E}\left(\int_0^L e^{\alpha \mathcal{Z}(t)-\beta t}\mathbf{1}_{\mathcal{J}(t)}dt\right)\Bigg|_{\alpha=0}, \tag{25}$$

where

$$\mathbf{E}\left(\int_0^L e^{\alpha \mathcal{Z}(t)-\beta t}\mathbf{1}_{\mathcal{J}(t)}dt\right) = \left[e^{\alpha Q}\mathbb{E}\left(e^{-\beta L}\right) - \mathbb{I} - \alpha\left(\mathbf{0}, \hat{\mathbf{L}}(\beta)\right)\right](\mathbb{K}(\alpha) - \beta\mathbb{I})^{-1}\mathbf{e}. \tag{26}$$

Proof. The proof and the derivation of $\hat{\mathbf{L}}(\beta)$ are given in Appendix A. ■

In order to finish the holding cost we have to find the $(n \times 1)$ vector $\mathbf{E}\left(\int_0^{T'} e^{-\beta t} I^-(t)dt\right)$. Recall that since we are now dealing with the OFF period, for all the matrices in **Table 1**, we have to add the index $(-)$. The OFF period stars at time L. We shift the time origin to L so that the OFF period starts at time 0. Consider $\tilde{\mathcal{X}}(t)$ similar to (9) but with $\mathbb{C}^-$ and $\tilde{\mathcal{X}}(0) = Q$, $\mathcal{J}(0) \in \mathfrak{J}$. Clearly, the latter process up to time T', i.e., $\left(\tilde{\mathcal{X}}(t)\right)_{0 \le t < T'}$, has the same distribution as $(I(t))_{0 \le t < T'}$. Similar arguments to (15) leads to:

$$\mathbf{E}\left(\int_0^{T'} e^{-\beta t} I(t)dt\right) = \frac{d}{d\alpha}\mathbf{E}\left(\int_0^{T'} e^{\alpha \tilde{\mathcal{X}}(t)-\beta t}dt\right)\Bigg|_{\alpha=0}, \tag{27}$$

where

$$\begin{aligned}\mathbf{E}\left(\int_0^{T'} e^{\alpha \tilde{\mathcal{X}}(t)-\beta t}dt\right) &= \left[\mathbb{E}\left(e^{\alpha \tilde{\mathcal{X}}(T')-\beta T'}\mathbf{1}_{\mathcal{J}(T')}\right) - \mathbb{E}\left(e^{\alpha \tilde{\mathcal{X}}(0)}\mathbf{1}_{\mathcal{J}(0)}\right)\right](\mathbb{K}(\alpha) - \beta\mathbb{I})^{-1}\mathbf{e} \\ &= \left(\mathbb{E}\left(e^{-\beta T'}\right) - e^{-\alpha Q}\mathbb{I}\right)(\mathbb{K}(\alpha) - \beta\mathbb{I})^{-1}\mathbf{e}.\end{aligned} \tag{28}$$

c. **Production cost.** Let q_i be the production cost for one unit. We have:

$$\begin{aligned}&PC(\beta) = \gamma\hat{\mathbf{P}}(\beta), \\ &\hat{\mathbf{P}}(\beta) = \left(\mathbb{I} - \mathbb{E}\left(e^{-\beta C}\right)\right)^{-1}\Delta_{\mathbf{q}}\mathbf{E}\left(\int_0^L p_{\mathcal{J}(t)}e^{-\beta t}dt\right).\end{aligned} \tag{29}$$

Note that the OFF period is characterized by no production. For the derivation of $\mathbf{E}\left(\int_0^L p_{\mathcal{J}(t)}e^{-\beta t}dt\right)$ (ON period), let $\mathbb{P}_j^+ = diag\{p_i, i \in \mathfrak{J}_j\}, j = 1, 2$ and $\mathbb{P}^+ = diag\left(\mathbb{P}_1^+, \mathbb{P}_2^+\right)$. Thus,

$$\mathbf{E}\left(\int_0^L p_{\mathcal{J}(t)}e^{-\beta t}dt\right) = \mathbb{E}\left(\int_0^L e^{-\beta t}\mathbf{1}_{\mathcal{J}(t)}dt\right)\mathbb{P}^+\mathbf{e}. \tag{30}$$

Applying (26) with $\alpha = 0$ yields

$$\mathbb{E}\left(\int_0^L e^{-\beta t}\mathbf{1}_{\mathcal{J}(t)}dt\right) = \left(\mathbb{E}\left(e^{-\beta L}\right) - \mathbb{I}\right)\left(\mathbb{K}(0) - \beta\mathbb{I}\right)^{-1}, \tag{31}$$

which **completes** the derivation of the production cost.

d. *Lost demand cost.* In our model, backlog is not allowed; any demand which cannot be satisfied immediately is lost. Clearly, there is no unsatisfied demand during the OFF period. During ON period, once level 0 is reached the process stays there until the environment changes to state with a positive growth rate. Assume the process hits 0 at state i (for some $c_i^+ < 0$), the demand is lost with rate $\left(-c_i^+\right)$ until the environmental state changes. Let wdt be the cost for a lost unit during a time interval of length dt $(w > 0)$. As a measure for the expected discounted lost demand cost one can use the functional

$$\begin{aligned} UC(\beta) &= \gamma\hat{\mathbf{U}}(\beta) \\ \hat{\mathbf{U}}(\beta) &= -w\gamma\left(\mathbb{I} - \mathbb{E}\left(e^{-\beta C}\right)\right)^{-1}\mathbf{E}\left(\int_0^L e^{-\beta t}c^+_{\mathcal{J}(t)}\mathbf{1}_{\{I^+(t)=0\}}dt\right) \end{aligned} \tag{32}$$

The right term is the expected discounted production loss during the first cycle; it can be written in terms of the local time process.

Corollary 4.1 *It is easy to verify that*

$$\mathbf{E}\left(\int_0^L e^{-\beta t}c^+_{\mathcal{J}(t)}\mathbf{1}_{\{I^+(t)=0\}}dt\right) = \mathbf{E}\left(\int_0^L e^{-\beta t}d\mathcal{L}(t)\right) = \hat{\mathbf{L}}(\beta)\mathbf{e}, \tag{33}$$

where $\hat{\mathbf{L}}(\beta)$ *is given in* (40).

A simple cost function for the entire system would be the sum, $TC(\beta)$, of these four expected discounted costs:

$$TC(\beta) = SC(\beta) + PC(\beta) + HC(\beta) + UC(\beta). \tag{34}$$

5. Summary

During the past few decades, the problem of control of inventory systems has been widely investigated. Many stochastic factors inherent in inventory systems can make it more difficult for managers to plan and control the inventory. Dealing with the randomness of demand, production, and returns, this study considers a continuous-review inventory system where the inventory level is characterized as a fluid process modelled by Markovian environment. The cost structure includes an order cost, a purchase cost, a set up cost, a production cost, an inventory cost, and a lost cost due to unsatisfied demands. By taking a simple probability approach and by applying stopping time theory to fluid processes and martingales, the explicit components of the resulting costs are derived. These cost components can be used for optimization purposes. Moreover, the closed-form expression of the components allows us to obtain efficiently and numerically the optimal parameters and enables us to investigate the behavior of the system and to study its properties. From a managerial perspective, our framework can be applied to many industries, in situations where the system is subject to uncertain environment. Our approach appears to be a powerful way to address related inventory problems. The models considered here assume only discrete demand and return sizes. For instance, it seems to be possible to adapt a similar approach in the case of continuous demand size distributions, such as exponential, uniform, or gamma. Another avenue of research would be to extend this framework to include random lead times; in that case, two policies can be applied while shortage, backordering or lost sales. All of these nontrivial extensions are tractable and worthy of study.

Appendix A.

Proof of claim 4.1. Theorem 2.1 of Asmussen and Kella [31] yields that the process

$$\begin{aligned}M(\alpha,t) &= \int_{v=0}^{t} e^{\alpha W(v)}\mathbf{1}_{\mathcal{J}(v)}dv\mathbb{K}(\alpha) + e^{\alpha W(0)}\mathbf{1}_{\mathcal{J}(0)} - e^{\alpha W(t)}\mathbf{1}_{\mathcal{J}(t)} + \alpha\left(\int_{v=0}^{t} e^{\alpha W(v)}\mathbf{1}_{\mathcal{J}(v)}dY(v)\right)\\ &= \int_{v=0}^{t} e^{\alpha\mathcal{Z}(v)-\beta v}\mathbf{1}_{\mathcal{J}(v)}dv(\mathbb{K}(\alpha)-\beta\mathbb{I}) + e^{\alpha\mathcal{Z}(0)}\mathbf{1}_{\mathcal{J}(0)} - e^{\alpha\mathcal{Z}(t)-\beta t}\mathbf{1}_{\mathcal{J}(t)}\\ &\quad+\alpha\left(\int_{v=0}^{t} e^{-\beta v}\mathbf{1}_{\mathcal{J}(v)}d\mathcal{L}(v)\right)\end{aligned} \tag{35}$$

is an n-row vector-valued zero mean martingale. The OST yields $\mathbb{E}(M(\alpha,L)) = \mathbb{E}(M(\alpha,0)) = 0$,

$$\begin{aligned}\mathbb{E}\left(\int_0^L e^{\alpha\mathcal{Z}(t)-\beta t}\mathbf{1}_{\mathcal{J}(t)}dt\right) &= \left[\mathbb{E}\left(e^{\alpha\mathcal{Z}(L)-\beta L}\mathbf{1}_{\mathcal{J}(L)}\right) - \mathbb{E}\left(e^{\alpha\mathcal{Z}(0)}\mathbf{1}_{\mathcal{J}(0)}\right)\right.\\ &\quad\left.-\alpha\mathbb{E}\left(\int_0^L e^{-\beta t}\mathbf{1}_{\mathcal{J}(t)}d\mathcal{L}(t)\right)\right](\mathbb{K}(\alpha)-\beta\mathbb{I})^{-1}.\end{aligned}$$

Since $\mathcal{Z}(0) = 0$ and $\mathcal{J}(0) \in \mathfrak{I}$, we obtain the $(n \times n)$ matrix $\mathbb{E}\left(e^{\alpha\mathcal{Z}(0)}\mathbf{1}_{\mathcal{J}(0)}\right) = \mathbb{I}$. Applying $\mathcal{Z}(L) = Q$ leads to

$$\mathbb{E}\left(e^{\alpha\mathcal{Z}(L)-\beta L}\mathbf{1}_{\mathcal{J}(L)}\right) = e^{\alpha Q}\mathbb{E}\left(e^{-\beta L}\right). \tag{36}$$

To finish, we have to derive the $(n \times n)$ matrix $\mathbb{E}\left(\int_0^L e^{-\beta t}\mathbf{1}_{\mathcal{J}(t)}d\mathcal{L}(t)\right)$, which is the expected discounted lost demand until time L. Notice that a loss occurs only during states in $\mathfrak{I}_2$. For that, we introduce the $(n_2 \times n_2)$ matrix $\Upsilon(\beta)$ whose ij_{th} component, $\Upsilon_{ij}(\beta)$, is the expected discounted loss in state $j \in \mathfrak{I}_2$ until exiting level 0, given that the process starts at level 0 with state $i \in \mathfrak{I}_2$.

Lemma A.1 $\Upsilon_{ij}(\beta)$ satisfies the following system of linear equations:

$$\begin{aligned}\Upsilon_{ii}(\beta) &= \frac{-c_i}{\beta}\left(1+\frac{\mathbb{Q}_{22}(i,i)}{\beta-\mathbb{Q}_{22}(i,i)}\right)+\sum_{j\neq i}\frac{\mathbb{Q}_{22}(i,j)}{\beta-\mathbb{Q}_{22}(i,i)}\Upsilon_{ji}(\beta),\\ \Upsilon_{ij}(\beta) &= \sum_{k\neq i}\frac{\mathbb{Q}_{22}(i,k)}{\beta-\mathbb{Q}_{22}(i,i)}\Upsilon_{kj}(\beta).\end{aligned} \tag{37}$$

Proof. Once the inventory process hits 0 in state $i \in \mathfrak{I}_2$, it stays there for an exponential random time ξ_i with parameter $-\mathbb{Q}_{22}(i,i)$. With probability $\frac{\mathbb{Q}_{22}(i,j)}{-\mathbb{Q}_{22}(i,i)}$ the state changes to $j \in \mathfrak{I}$, and the expected discounted loss from state j is $\Upsilon_j(i)$. By conditioning on the first state visited after i, we readily obtain that

$$\begin{aligned}\Upsilon_{ii}(\beta) &= -E\left(\int_{t=0}^{\xi_i} c_i e^{-\beta t}dt\right)+E_i\left(e^{-\beta\xi_i}\right)\sum_{\substack{i\neq j\\ j\in\mathfrak{I}_2}}\frac{\mathbb{Q}_{22}(i,j)}{-\mathbb{Q}_{22}(i,i)}\Upsilon_{ji}(\beta),\\ \Upsilon_{ij}(\beta) &= E_i\left(e^{-\beta\xi_i}\right)\sum_{\substack{k\neq i\\ k\in\mathfrak{I}_2}}\frac{\mathbb{Q}_{22}(i,k)}{-\mathbb{Q}_{22}(i,i)}\Upsilon_{kj}(\beta).\end{aligned} \tag{38}$$

Solving (38) with respect to $\Upsilon_{ii}(\beta)$ and $\Upsilon_{ij}(\beta)$ returns (37) (note that $E\left(\int_{t=0}^{\xi_i} c_i e^{-\beta t}dt\right) = \frac{c_i}{\beta}\left(1+\frac{\mathbb{Q}_{22}(i,i)}{\beta-\mathbb{Q}_{22}(i,i)}\right)$ and $E_i\left(e^{-\beta\xi_i}\right) = \frac{-\mathbb{Q}_{22}(i,i)}{\beta-\mathbb{Q}_{22}(i,i)}$)). ■.

Now, we apply $\Upsilon(\beta)$ to the derivation of the loss until time L. Since the lost demand occurs only for states in $\mathfrak{I}_2$, the matrix $\mathbb{E}\left(\int_{v=0}^{L} e^{-\beta v}\mathbf{1}_{\mathcal{J}(v)}d\mathcal{L}(v)\right)$ has the form

$$\mathbb{E}\left(\int_0^L e^{-\beta t}\mathbf{1}_{\mathcal{J}(t)}d\mathcal{L}(t)\right) = \left(\mathbf{0},\hat{\mathbf{L}}(\beta)\right) \tag{39}$$

where $\hat{\mathbf{L}}(\beta)$ is an $(n \times n_2)$ matrix whose ij_{th} component is the expected discounted loss in state $j \in \mathfrak{I}_2$ until L, given $\mathcal{J}(0) = i \in \mathfrak{I}$. Let $\hat{\mathbf{L}}_i(\beta), i = 1,2$ be an $(n_i \times n_2)$ sub-matrix of $\hat{\mathbf{L}}(\beta)$ includes all rows corresponding to states in $\mathfrak{I}_i, i = 1,2,$ such that

$$\hat{\mathbf{L}}(\beta) = \begin{pmatrix}\hat{\mathbf{L}}_1(\beta)\\ \hat{\mathbf{L}}_2(\beta)\end{pmatrix}. \tag{40}$$

It is easy to verify that

$$\begin{aligned}\hat{\mathbf{L}}_1(\beta) &= {}^{Q}\Psi(\beta)\left(\Upsilon(\beta)+(\beta\mathbb{I}-\mathbb{Q}_{22})^{-1}\mathbb{Q}_{21}\hat{\mathbf{L}}_1(\beta)\right),\\ \hat{\mathbf{L}}_2(\beta) &= \Upsilon(\beta)+(\beta\mathbb{I}-\mathbb{Q}_{22})^{-1}\mathbb{Q}_{21}\hat{\mathbf{L}}_1(\beta).\end{aligned} \tag{41}$$

Lost demand occurs when the process drops to level 0 avoiding Q, with LST ${}^{Q}\Psi(\beta)$. From that point, $\Upsilon(\beta)$ is the discounted lost demand. The term $(\beta\mathbb{I} - \mathbb{Q}_{22})^{-1}\mathbb{Q}_{21}$ is the discounted time until exiting level 0 with environmental state in $\mathfrak{I}_1$ and starting again with LST $\hat{\mathbf{L}}_1(\beta)$. The matrix $\hat{\mathbf{L}}_2(\beta)$ is derived similarly.

Author details

Yonit Barron
Ariel University, Ariel, Israel

*Address all correspondence to: ybarron@ariel.ac.il

References

[1] Nahmias S. Production and Operations Analysis. 3rd edition. New York: Irwin/McGraw-Hill; 1997

[2] Beltran JL, Krass D. Dynamic lot sizing with returning items and disposals. IIE Transactions. 2002;**34**:437-448

[3] Fleischmann M, Kuik R, Dekker R. Controlling inventories with stochastic item returns: A basic model. European Journal of Operational Research. 2002; **138**(1):63-75

[4] Pinçe Ç, Gürler Ü, Berk E. A continuous review replenishment-disposal policy for an inventory system with autonomous supply and fixed disposal costs. European Journal of Operational Research. 2008;**190**:421-442

[5] Shaharudin MR, Zailani S, Tan KC. Barriers to product returns and recovery management in a developing country: investigation using multiple methods. Journal of Cleaner Production. 2013;**96**: 220-232

[6] Barron Y. An order-revenue inventory model with returns and sudden obsolescence. Operations Research Letters. 2018;**46**(1):88-92

[7] Barron Y. Performance analysis of a reflected fluid production/inventory model. Mathematical Methods of Operations Research. 2016;**83**(2016):1-31

[8] Boxma OJ, Kella O, Perry D. An intermittent fluid system with exponential on-times and semi-Markov input rates. Probability in the Engineering and Information Sciences. 2001;**15**:189-198

[9] Kulkarni VG, Yan K. A fluid model with upward jumps at the boundary. Queueing Systems. 2007;**56**:103-117

[10] Bean N, O'reilly M. Performance measure of a multi-layer Markovian fluid model. Annals of Operations Research. 2008;**160**:99-120

[11] Barron Y, Hermel D. Shortage decision policies for a fluid production model with MAP arrivals. International Journal of Production Research. 2017;**55** (14):3946-3969

[12] Baek JW, Lee HW, Lee SW, Ahn S. A Markov modulated fluid flow queueing model under D-policy. Numerical Linear Algebra with Applications. 2011;**18**:993-1010

[13] Baek JW, Ahn S. Analyses of the Markov modulated fluid flow with one-sided ph-type jumps using coupled queues and the completed graphs. Journal of the Korean Statistical Society. 2014;**43**(2014):415-424

[14] Kulkarni VG, Yan K. Production-inventory systems in stochastic environment and stochastic lead times. Queueing Systems. 2012;**70**:207-231

[15] Barron Y. A fluid EOQ model with Markovian environment. Journal of Applied Probability. 2015;**52**(2):473-489

[16] Mohebbi E. A production-inventory model with randomly changing environmental conditions. European Journal of Operational Research. 2006; **174**:539-552

[17] Kouki C, Legros B, Babai MZ, Jouini O. Analysis of base-stock perishable inventory systems with general lifetime and lead-time. European Journal of Operational Research. 2020;**287**(3):901-915

[18] Barron Y. Critical level policy for a production-inventory model with lost

sales. International Journal of Production Research. 2019A;**57**(6):1685-1705

[19] Barron Y, Dreyfuss M. A triple Ssℓ-thresholds base-stock policy subject to uncertainty environment, returns and order cancellations. Computers & Operations Research. 2021;**134**:105320

[20] Yan K, Kulkarni VG. Optimal inventory policies under stochastic production and demand rates. Stochastic Models. 2008;**24**:173-190

[21] Kulkarni VG. Modeling and analysis of stochastic systems. 2nd edition. Boca Raton: CRC Press; 2009

[22] Berman O, Perry D, Stadje W. A fluid EOQ model with a two-state Random environment. Probability in the Engineering and Informating Sciences. 2006;**20**:329-349

[23] Berman O, Perry D, Stadje W. Optimal replenishment in a Brownian motion EOQ model with hysteretic parameter changes. International Journal of Inventory Research. 2008;**1**:1-18

[24] Berman O, Perry D. An EOQ model with state dependent demand rate. European Journal of Operational Research. 2006;**176**:255-272

[25] Kella O, Perry D, Stadje W. A stochastic clearing model with Brownian and a compound Poisson component. Probability in the Engineering and Informational Sciences. 2003;**17**:1-22

[26] Berman O, Parlar M, Perry D, Posner MJM. Production/Clearing models under continuous and sporadic review. Methodology and Computing in Applied Probability. 2005;**7**:203-224

[27] Barron Y. Clearing control policies for MAP inventory process with lost sales. European Journal of Operational Research. 2016;**251**(2):495-508

[28] Vickson RG. A single product cycling problem under Brownian motion demand. Management Science. 1986;**32**:1336-1345

[29] Kella O, Whitt W. A Storage model with a two-stage random environment. Operations Research. 1992;**40**:s257-s262

[30] Boxma O, Kaspi H, Kella O, Perry D. On/off storage systems with state-dependent input, output, and switching rates. Probability in the Engineering and Informational Sciences. 2005;**19**:1-14

[31] Asmussen S, Kella O. A multi-dimensional martingale for Markov additive process and its applications. Advances in Applied Probability. 2000;**32**:376-393

[32] Barron Y. An skS fluid inventory model with exponential leadtimes and order cancellations. Stochastic Models. 2016;**32**:301-332

[33] Ramaswami V. Passage times in fluid models with application to risk processes. Methodology and Computing in Applied Probability. 2006;**8**:497-515

[34] Ahn S, Badescu AL, Ramaswami V. Time dependent analysis of finite buffer fluid flows and risk models with a dividend barrier. Queueing Systems. 2007;**55**:207-222

[35] Ahn S, Ramaswami V. Fluid flow models and queues- a connection by stochastic coupling. Stochastic Models. 2003;**19**:325-348

[36] Ahn S, Ramaswami V. Efficient algorithms for transient analysis of stochastic fluid flow models. Journal of Applied Probability. 2005;**42**:531-549

[37] Ahn S, Ramaswami V. Transient analysis of fluid models via elementary

level-crossing arguments. Stochastic Models. 2006;**22**:129-147

[38] Bean N, O'reilly M, Taylor P. Algorithms for the Laplace-Stieltjes Transforms of First Return Times for Stochastic Fluid Flows. Methodology and Computing in Applied Probability. 2008; **10**:381-408

[39] Doob J. Stochastic Processes. New York: Wiley; 1953

[40] Ross M. Stochastic Processes. 2nd edition. New York: Wiley; 1996

[41] Asmussen S. Applied Probability and Queues. Second edition. New York, Inc: Springer-Verlag; 2003